INTERNATIONAL CENTRE FOR MECHANICAL SCIENCES

COURSES AND LECTURES - No. 298

CHAOTIC MOTIONS IN NONLINEAR DYNAMICAL SYSTEMS

W. SZEMPLIŃSKA-STUPNICKA
POLISH ACADEMY OF SCIENCES

G. IOOSS
UNIVERSITÉ DE NICE

F.C. MOON
CORNELL UNIVERSITY

SPRINGER-VERLAG WIEN GMBH

Le spese di stampa di questo volume sono in parte coperte da contributi
del Consiglio Nazionale delle Ricerche.

This volume contains 88 illustrations.

Originally published by Springer Verlag Wien-New York in 1988

ISBN 978-3-211-82062-9 ISBN 978-3-7091-2596-0 (eBook)
DOI 10.1007/978-3-7091-2596-0

PREFACE

The discovery of new types of dynamic behavior in physical systems in the last decade has brought about new analytic and experimental techniques in dynamics. Principal amongst these new discoveries is the existence of chaotic, unpredictable behavior in many nonlinear deterministic systems. Observations of chaotic and prechaotic behavior have been observed in all areas of classical physics including solid and fluid mechanics, thermo-fluid phenomena, electromagnetic systems and in the area of acoustics.

The lectures presented in this book, look at the field of chaotic and nonlinear dynamics from three different points of view. In the first set of lectures F. Moon outlines many of the new experimental techniques which have emerged from the study of chaotic vibrations. These include Poincaré sections, fractal dimensions and Lyaponov exponents. In the text by W. Szemplinska-Stupnicka, the relation between the new chaotic phenomena and classical perturbation techniques of nonlinear vibration is explored for the first time. Chaotic phenomena is often preceded by a series of bifurcations including subharmonic and limit cycle or Hopf bifurcations. Finally, in the third set of lecture notes G. Iooss presents methods of analysis for the calculation of bifurcation in nonlinear systems based on molecular geometric mathematical concepts.

The modern study of nonlinear dynamics is unique in the field of applied mathematics since it requires the analyst to become familiar with experiments (at least numerical ones) since chaotic solutions cannot be written down, and it requires the experimentalist to master the next concepts in the theory of nonlinear dynamical systems. This book is unique in that it presents both the viewpoint of the analyst and the experimenter in nonlinear and chaotic dynamics. It should be of interest to engineers, physicists and applied mathematicians interested in chaotic and stochastic phenomena.

It is indeed fitting that these lectures were given in Udine, Italy near where Galileo made his earlier discoveries in dynamics. It is exciting to observe that often many centuries ago new studies in dynamics were being made that will have an important impact in both classical physics and applied science.

Wanda Szemplińska-Stupnicka

CONTENTS

EXPERIMENTS IN CHAOTIC DYNAMICS*

F. Moon
Cornell University, Ithaca, New York, USA

ABSTRACT

The discovery of deterministic chaotic vibrations in nonlinear dynamical systems has led to new mathematical ideas and analytical techniques in nonlinear dynamics. Along with these new ideas has come a host of new experimental tools to analyze vibrations in physical systems including Poincaré maps, bifurcation diagrams, chaos criteria diagrams, Lyapunov exponents and fractal dimensions. Some of these new experimental methods are reviewed in these notes, particularly as they apply to nonlinear mechanical systems.

---

* A summary of lectures given at the International Center for Mechanical Sciences (CISM) at Udine, Italy, July 1986.

1. INTRODUCTION: NEW TOOLS TO DIAGNOSE CHAOTIC VIBRATIONS

The modern advances in nonlinear dynamics in both mathematical theory and analytical techniques have been matched by the development of new experimental and numerical tools for studying the dynamics of nonlinear systems. For this summary we shall review these new experimental tools. These notes were based in part on the book by the author, Chaotic Vibrations, published by J. Wiley and Sons, New York, 1987.

The list of experimental and numerical tools to study systems with chaotic dynamics includes the following:

Phase plane methods
Pseudo phase plane methods
Bifurcation diagrams
Fast Fourier Transforms
Auto-correlation functions
Poincaré maps
Double Poincaré maps
Reduction to one-dimensional maps
Lyapunov exponents
Fractal dimensions
Invariant distributions
Chaos diagrams
Basin boundary diagrams.

These techniques expand the engineer's ability to analyze the dynamics of complex systems. The specific technique depends in part on the particular nature of the system. However, in the case of simple one-degree-of-freedom nonlinear systems, a simple procedure has been developed to look for chaotic behavior.

Engineers often have to diagnose the source of unwanted oscillations in physical systems. The ability to classify the nature of oscillations can provide a clue as to how to control them. For example, if the system is thought to be *linear*, large periodic oscillations may be traced to a *resonance* effect. However, if the system is *nonlinear*, a *limit cycle* may be the source of periodic vibration, which in turn may be traced to some *dynamic instability* in the system.

In order to identify nonperiodic or chaotic motions, the following checklist is provided:

i) Identify nonlinear elements in the system
ii) Check for sources of random input in the system
iii) Observe time history of measured signal
iv) Look at phase plane history
v) Examine Fourier spectrum of signal
vi) Take Poincaré map of signal
vii) Vary system parameters (bifurcation diagram)

One can also measure two properties of the motion:

a) Fractal dimension
b) Lyapunov exponents

To focus the discussion, I will use the vibration of the buckled beam as an example to illustrate the characteristics of chaotic dynamics.

1.1 Nonlinear elements

A chaotic system must have nonlinear elements or properties. *A linear system cannot exhibit chaotic vibrations.* In a linear system periodic inputs produce periodic outputs of the same period (Figure 1). In mechanical systems nonlinear effects include:

a) nonlinear elastic or spring elements
b) nonlinear damping, such as stick-slip friction
c) backlash, play or bilinear springs
d) most systems with fluids
e) nonlinear boundary conditions

Nonlinear elastic effects can reside in either material properties or geometric effects. For example, the relation between stress and strain in rubber is nonlinear. However, while the stress-strain law for steel is usually linear below yield, large displacement bending of a beam, plate or shell may exhibit nonlinear relations between the applied forces or moments and displacements. Nonlinearities due to large displacements or rotations are usually called geometric nonlinearities.

In the example of the buckled beam, identification of the nonlinear element is easy (Figure 2). Any mechanical device that has more than one static equilibrium position either has play, backlash or nonlinear stiffness. In the case of the beam buckled by end loads (Figure 2a) the geometric nonlinear stiffness is the culprit. If the beam is buckled by magnetic forces (Figure 2b) the nonlinear magnetic forces are the sources of chaos in the system.

1.2 Random inputs

By definition, chaotic vibrations arise from deterministic physical systems or deterministic differential or difference equations. While noise is always present in experiments, and even numerical simulations it is presumed that large nonperiodic signals do not arise from very small input noise. Thus a very low input noise is required if one is to attribute nonperiodic response to a deterministic system behavior.

1.3 Time history

Usually the first clue that the experiment has chaotic vibrations is the observation of the signal amplitude with time on a chart recorder or oscilloscope (Figure 3). The motion is observed to exhibit no visible pattern or periodicity. This test is not foolproof, however, since a motion could have a long period behavior that is not easily detected. Also, some nonlinear systems exhibit quasi-periodic vibrations when two incommensurate periodic signals are present. Here the signal may appear to be nonperiodic, but it can be broken down into the sum of two or more periodic signals.

1.4 Phase plane

Consider a one-degree-of-freedom mass with displacement x(t) and velocity v(t). Its equation of motion from Newton's law can be written in the form

$$\dot{x} = v$$

$$\dot{x} = \frac{1}{m} f(x,v,t) \qquad (1)$$

where m is the mass, and f is the applied force. The phase plane is defined as the set of points (x,v) (some authors use the momentum mv instead of v). When the motion is periodic (Figure 4a), the phase plane orbit traces out a closed curve which is best observed on an analog oscilloscope. For example, the forced oscillations of a linear spring-mass-dashpot system exhibits an elliptically shaped orbit. However, a forced nonlinear system with a cubic spring element may show an orbit which crosses itself but is still closed. This can represent a subharmonic oscillation.

Chaotic motions on the other hand have orbits which never close or repeat. Thus the trajectory of the orbits in the phase plane will tend to fill up a section of the phase space as in Figure 4b. Although this wandering of orbits is a clue to chaos, continuous phase plane plots provide very little information and one must use a modified phase plane technique called Poincaré maps.

Often one has only measured a single variable v(t). If v(t) is a velocity variable, then one can integrate to get x(t) so that the phase plane consists of points

$$\left[\int_0^t v\,d\tau\ ,\ v(t)\right]. \tag{2}$$

On the other hand, if one has to differentiate a displacement or strain related signal x(t), high frequency noise is often introduced. In this case the experimenter is advised to use a good low pass filter on x(t) before differentiation.

Another technique that has been used to obtain phase plane orbits when only one variable is measured is the time delayed pseudo-phase plane method. For a one-degree-of-freedom system with measurement x(t), one plots the signal versus itself but delayed or advanced by a fixed time constant [x(t), x(t+T)]. The idea here is that the signal x(t+T) is related to $\dot{x}(t)$ and should have properties similar to those in the classic phase plane $[x(t),\dot{x}(t)]$. In Figure 5 we show a pseudo-phase plane orbit for a harmonic oscillator. If the motion is chaotic the trajectories do not close (Figure 6). The choice of T is not crucial, except to avoid a natural period of the system.

1.5 Fourier spectrum

One of the clues to detecting chaotic vibration is the appearance of a broad spectrum of frequencies in the output when the input is a single frequency harmonic motion, or is DC (Figure 7). This characteristic of chaos becomes more important if the system is of low dimension (i.e., one to three degrees of freedom). Often there is an initial dominant frequency component ω_0, a precursor to chaos is the appearance of subharmonics in the frequency spectrum, ω_0/n. In addition to ω_0/n, harmonics of this frequency will also be present of the form $m\omega_0/n$ (m,n = 1,2,3,...).

One must be cautioned, though, against concluding that multiharmonic outputs imply chaotic vibrations since the system in question might have many hidden degrees of freedom of which the observer is not aware. In large degree of freedom systems the use of the Fourier spectrum may not be of much help in detecting chaotic vibrations unless one can observe changes in the spectrum as some parameter is varied such as the driving amplitude or frequency.

1.6 Poincaré maps

In the mathematical study of dynamical systems a *map* refers to a time sampled sequence of data $\{x(t_1), x(t_2),...,x(t_n),...,x(t_N)\}$ with the notation $x_n \equiv x(t_n)$. A deterministic map is one in which the values of x_i for $i \geq n+1$ can be determined from the values of x_i, $i \leq n$. In the simplest case this is written in the form

$$x_{n+1} = f(x_n)\,. \tag{3}$$

This can be recognized as a *difference equation.* The idea of a map can be generalized to more than one variable.

For example, suppose we consider the motion of a particle as displayed in the phase plane $(x(t), \dot{x}(t))$. We learned above that when the motion is chaotic the trajectory tends to fill up a portion of phase space. However, if instead of looking at the motion continuously, we only *look at the motion at discrete times.* Then the motion will appear as a sequence of dots in the phase plane (Figure 8). If $x_n \equiv x(t_n)$, $y_n \equiv \dot{x}(t_n))$ then this sequence of points in the phase plane represents a *two dimensional map*

$$\begin{aligned} x_n &= f(x_n, y_n) \\ y_n &= g(x_n, y_n)\,. \end{aligned} \tag{4}$$

When the sampling times t_n are chosen according to certain rules, to be discussed below, this map is called a *Poincaré Map.*

1.6.1 Poincaré maps for forced vibration systems

When there is a driving motion of period T, a natural sampling rule for a Poincaré map is to choose $t_n = nT + \tau_0$. This allows one to distinguish between periodic motions and nonperiodic motions. For example, if we sample the harmonic motion shown in Figure 3, in synchrony with its period, its "map" in the phase plane will become a point. If the output, however, were a subharmonic of period three, then the Poincaré map would consist of a set of three points as shown in Figure 9. (In the jargon of mathematical dynamics, one says that the functions f( ), g( ) in (4) have three *fixed points.*)

Another nonchaotic Poincaré map occurs where the motion consists of two *incommensurate* frequencies

$$x(t) = C_1 \sin(\omega_1 t + d_1) + C_2 \sin(\omega_2 t + d_2) \tag{5}$$

where ω_1/ω_2 is an irrational number. If one samples at a period corresponding to either frequency, the map in the phase plane will become a continuous *closed figure* or orbit. This motion is sometimes called *almost periodic* or *quasi-periodic* motion or motion on a torus and is not considered to be chaotic.

Finally, if the Poincaré map does not consist of either a finite set of points (Figure 9) or a closed orbit then the motion may be chaotic (Figure 10).

Here we must distinguish between damped and undamped systems. In undamped or lightly damped systems the Poincaré map of chaotic motions often appear as a cloud of unorganized points in the phase plane. Such motions are sometimes called *stochastic* (see, e.g., Lichtenberg and Lieberman (1983)). In damped systems the Poincaré map will sometimes appear as an infinite set of highly organized points arranged in what appear to be parallel lines as shown in Figure 10b. In numerical simulations one can enlarge a portion of the Poincaré map and see further structure. If this structured set of points continues to exist after several enlargements one says the motion behaves as a *strange attractor*. This embedding of structure within structure is often referred to as a *fractal set*.

The appearance of fractal patterns in the Poincaré map of a vibration history is a strong indicator of chaotic motions. For an autonomous system in a third order system one can define a Poincaré map by constructing a two dimensional oriented surface in this space and looking at the points (x_n, y_n, z_n) where the trajectory pierces this surface.

1.7 Bifurcation diagrams

1.7.1 Routes to chaos

In conducting any of these tests for chaotic vibrations one should try to vary one or more of the control parameters in the system. For example, in the case of the buckled structure (Figure 2) one can vary either the forcing amplitude or forcing frequency.

In changing a parameter one looks for a pattern of periodic responses. One characteristic precursor to chaotic motion is the appearance of subharmonic periodic vibrations. There may in fact be many patterns of prechaos behavior. However, several models of prechaotic behavior have been observed in both numerical and physical experiments.

1.7.2 Period doubling

In the period doubling phenomenon, one starts with a system with a fundamental periodic motion. Then as some experimental parameter is varied, say λ, the motion undergoes a bifurcation or change to a periodic motion with twice the period of the original oscillation. As λ is changed further, the system undergoes bifurcations to periodic motions with twice the period of the previous oscillation (Figure 11). One outstanding feature of this scenario is that the critical values of λ at which successive period doublings occur obey the following scaling rule:

$$\frac{\lambda_n - \lambda_{n-1}}{\lambda_{n+1} - \lambda_n} \rightarrow \delta = 4.6692016 \qquad (6)$$

as $n \rightarrow \infty$. (This is called the Feigenbaum number after the physicist who discovered this scaling behavior.) In practice, this limit approaches δ by the third or fourth bifurcation.

This process will accumulate at a critical value of the parameter after which the motion becomes chaotic.

This phenomenon has been observed in a number of physical systems as well as numerical simulations. This most elementary mathematical equation which illustrates this behavior is a first order difference equation,

$$x_{n+1} = 4\lambda x_n(1 - x_n) \tag{7}$$

As the system parameter is changed beyond the critical value, chaotic motions exist in a band of parameter values. However, these bands may be of finite width. That is, as the parameter is varied periodic windows may develop. Periodic motions in this regime may again undergo period doubling bifurcations again leading to chaotic motions.

1.7.3 Quasi-periodic route to chaos

While period doubling is the most celebrated scenario for chaotic vibration, there are several other schemes that have been studied and observed. In one proposed by Newhouse, Ruelle, and Takens (1978), they imagine a system which undergoes successive dynamic instabilities before chaos. For example, suppose a system is initially in a steady state and becomes dynamically unstable after changing some parameter (e.g. flutter). As the motion grows, nonlinearities come into effect, and the motion becomes a limit cycle. If after further parameter changes the system undergoes two more Hopf bifurcations so that three simultaneous coupled limit cycles are present, chaotic motions becomes possible.

Thus the precursor to such chaotic motion is the presence of two simultaneous periodic oscillations. When the frequencies of these oscillations ω_1, ω_2 are not commensurate, the observed motion itself is not periodic, but is said to be *quasi-periodic.* As discussed above, the Poincaré map of a quasi-periodic motion is a closed curve in the phase plane. Such motions are imagined to take place on the surface of a torus where the Poincaré map represents a plane which cuts the torus. If ω_1, ω_2 are incommensurate, the trajectories fill the surface of the torus. If ω_1/ω_2 is a rational number, the trajectory on the torus will eventually close though it might perform many orbits in both angular directions of the torus before closing. In the latter case the Poincaré map will become a set of points generally arranged in a circle. Chaotic motions are often characterized in such systems by the break-up of the quasi-periodic torus structure as the system parameter is varied.

Evidence for the three-frequency transition to chaos has been observed in flow between rotating cylinders (Couette flow) where vortices form with changes in the rotation speed.

1.7.4 Intermittancy

In a third route to chaos, one observes long periods of periodic motion with bursts of chaos. This scenario is called *intermittancy.* As one varies a parameter the chaotic bursts become more frequent and longer. Evidence for this model for prechaos has been claimed in experiments on convection in a cell (closed box) with a temperature gradient (called Rayleigh-Benard convection).

It should be noted that for some physical systems one may observe all three patterns of pre-chaotic oscillations and many more depending on the parameters of the problem. The value of identifying a particular pre-chaos pattern of motion with one of these now "classic" models is that a body of mathematical work exists on each which may offer better understanding of the chaotic physical phenomenon under study.

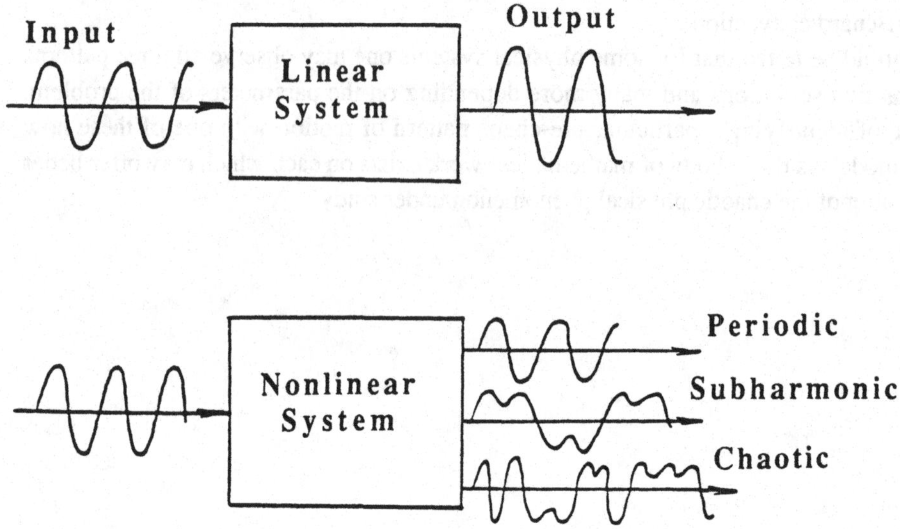

Figure 1.

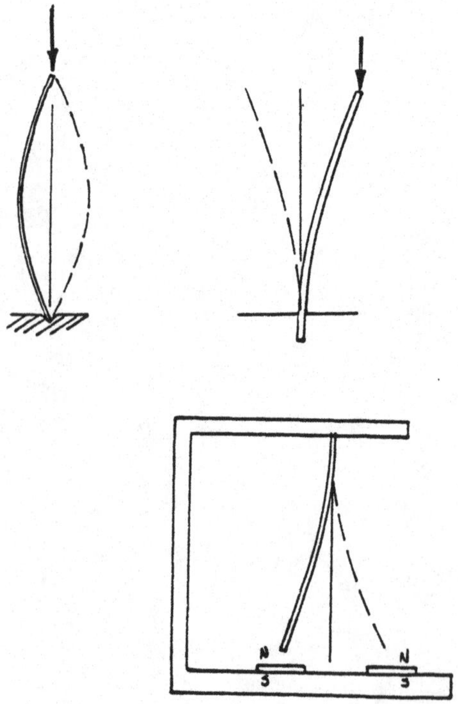

Figure 2.

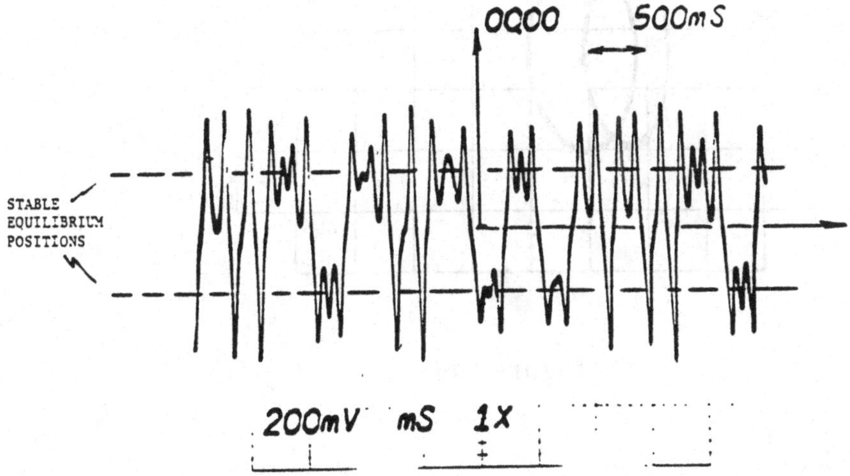

Figure 3.

Bending strain versus time for forced chaotic vibrations of the buckled beam.

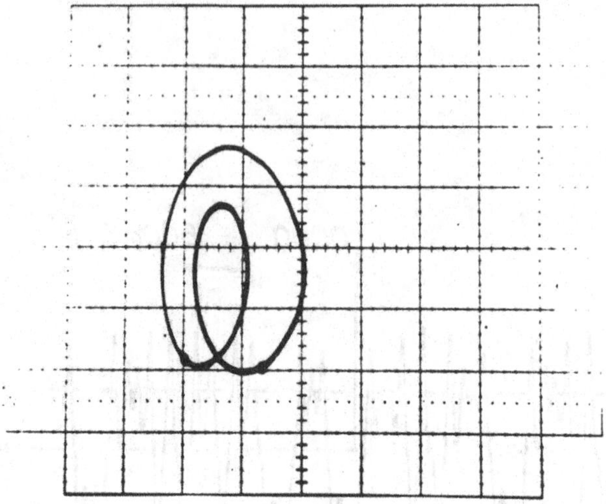

Figure 4a.

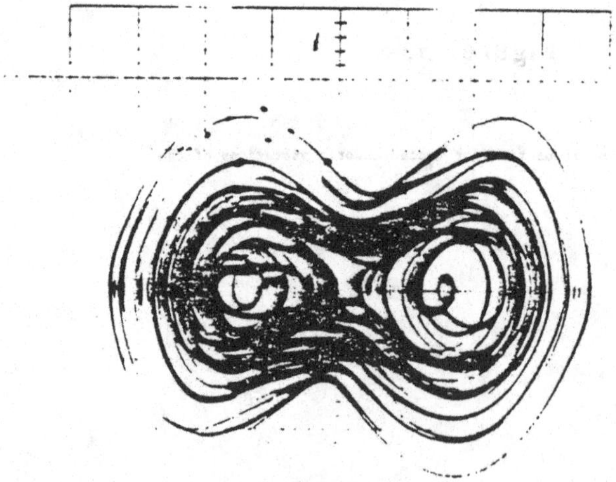

Figure 4b.

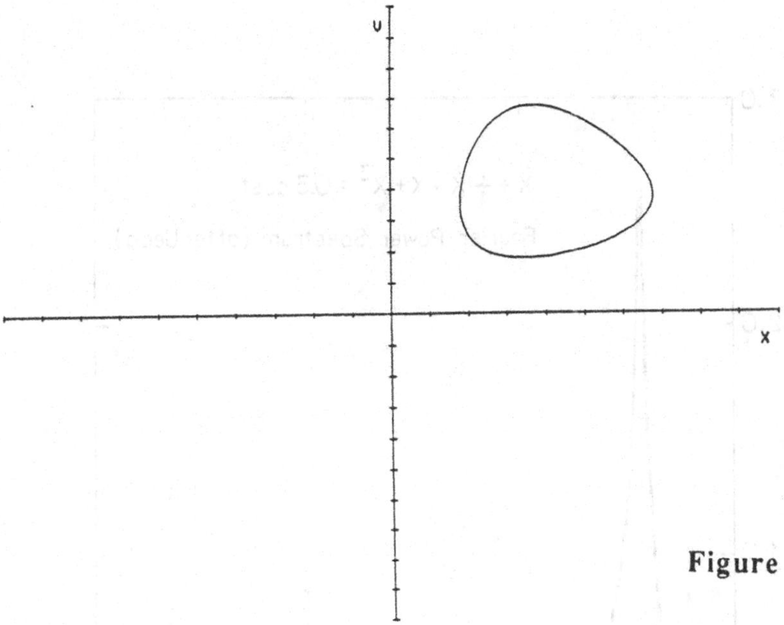

Figure 5.

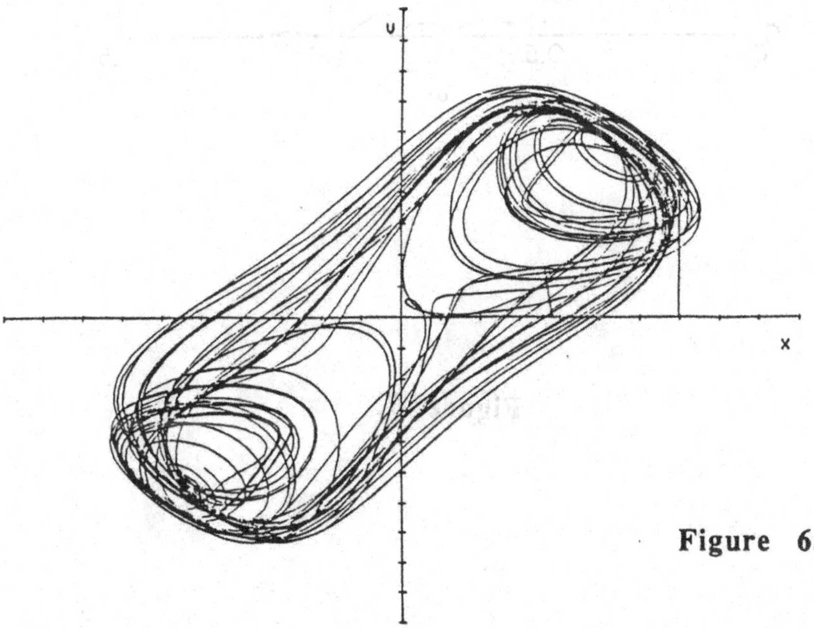

Figure 6.

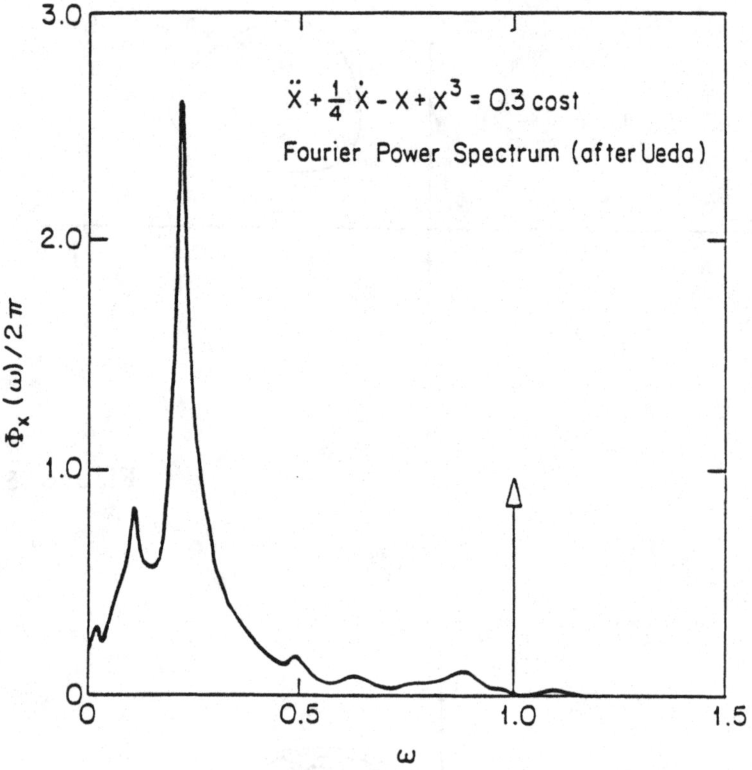

3.0
2.0
1.0
0
0.5
1.0
1.5
$\ddot{X} + \frac{1}{4}\dot{X} - X + X^3 = 0.3 \cos t$
Fourier Power Spectrum (after Ueda)
$\Phi_x(\omega)/2\pi$
ω

Figure 7.

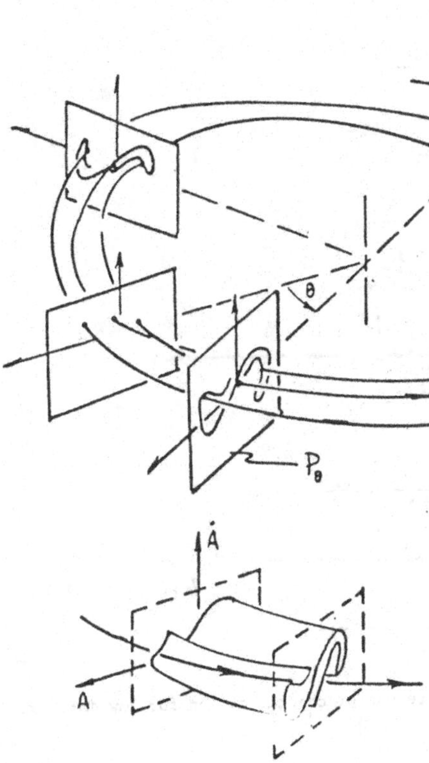

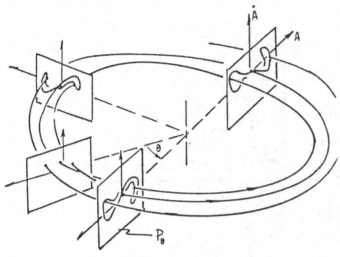

Sketch of strange attractor surfaces in the product space of Poincare plane and forcing amplitude phase.

Figure 8.

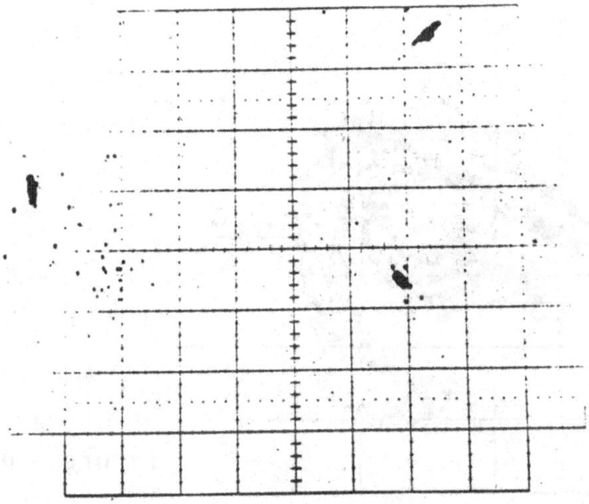

Figure 9.

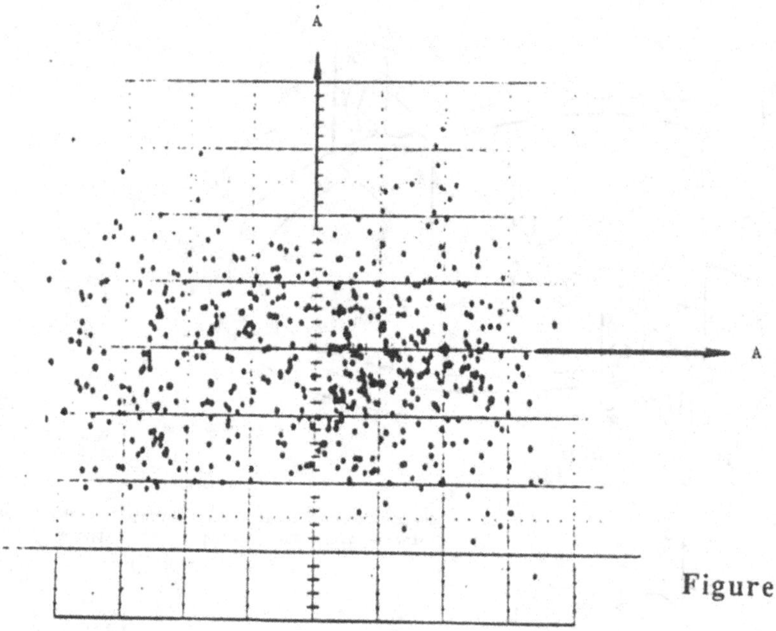

Figure 10a.

Experimental Poincare map of chaotic motion for low damping.

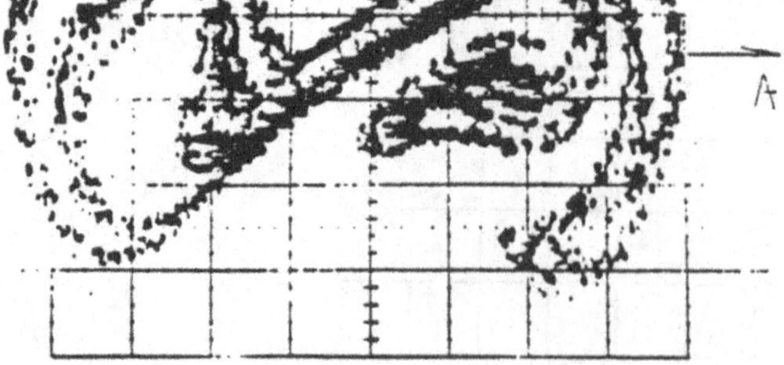

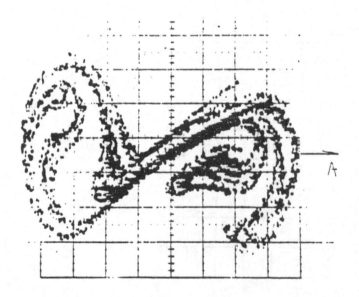

Figure 10b.

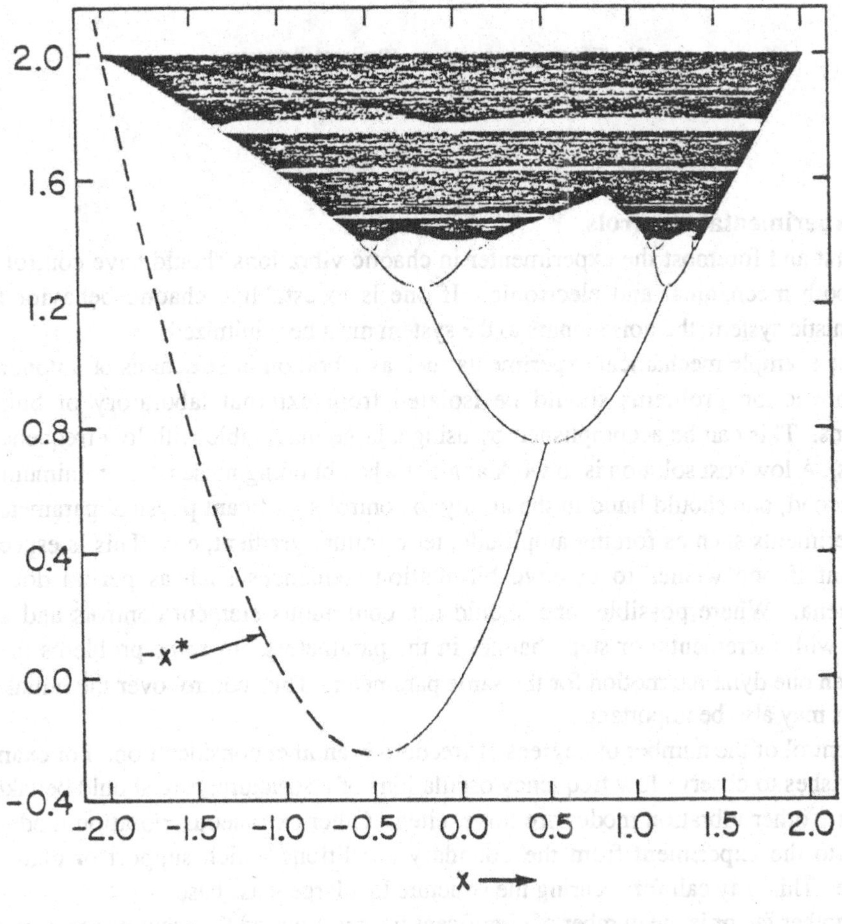

Bifurcation diagram for equation: $x_{n+1} = C - x_n^2$

From Grebogi, Ott & Yorke (1982)

Figure 11.

2. Experimental Techniques

2.1 Experimental controls

First and foremost the experimenter in chaotic vibrations should have control over noise, both mechanical and electronic. If one is to establish chaotic behavior for a deterministic system, the noise inputs to the system must be minimized.

For example mechanical experiments such as vibration of structures or autonomous fluid convection problems should be isolated from external laboratory or building vibrations. This can be accomplished by using a large mass table with low frequency air bearings. A low cost solution is to work at night when building noise is at a minimum.

Second, one should build in the ability to control significant physical parameters in the experiments such as forcing amplitude, temperature gradient, etc. This is especially important if one wishes to observe bifurcation sequences such as period doubling phenomena. Where possible, one should use continuous element controls and avoid devices with incremental or step changes in the parameters. In some problems there is more than one dynamic motion for the same parameter. Thus control over the initial state variables may also be important.

Control of the number of degrees of freedom is another consideration. For example, if one wishes to observe low frequency oscillations of a structure, care should be taken to make sure other vibration modes are not excited. Other extraneous vibration modes can creep into the experiment from the boundary conditions which support or clamp the structure. This may call for securing the structure to a large mass base.

Another factor is the number of significant figures required for accurate measurement. For example, to plot Poincaré maps from digitally sampled data, an 8 bit system may not be sensitive enough and one may have to go with 12 bit electronics or better. In some of our experiments on Poincaré maps, we have obtained better results from analog devices such as a good analog storage oscilloscope than by using an 8 bit digital oscilloscope, especially as regards resolution of fine fractal structure in the maps.

2.2 Frequency bandwidth

Most experiments in fluid, solid, or reacting systems may be viewed as infinite dimensional continua. However, one often tries to develop a mathematical model with a few degrees of freedom to explain the major features of the chaotic or turbulent motions of the system. One does this by not only making measurements at a few spatial locations in the continuous system, but by limiting the frequency bandwidth over which one observes the chaos. This is especially important if velocity measurements for phase plane plots are to be made from deformation histories. Electronic differentiation will amplify higher frequency signals, which may not be of interest in the experiment. Thus extremely good electronic filters are often required, especially ones that have little or no phase shift in the frequency band of interest.

2.3 Experimental Poincaré maps

Poincaré maps are one of the principal ways of recognizing chaotic vibrations in a low degree of freedom problem. We recall that the dynamics of a one-degree-of-freedom forced mechanical oscillator may be described in a three-dimensional phase space. Thus if $x(t)$ is the displacement, $(x,x,\omega t)$ represents a point in a cylindrical phase space where $\phi = \omega t$ represents the phase of the periodic forcing function. A Poincaré map for this problem consists of digitally sampled points in this three-dimensional space, e.g. $(x(t_n), \dot{x}(t_n), \omega t_n = 2\pi n)$. This map can be thought of as slicing a torus.

Experimentally this can be done in several ways. If one has a storage oscilloscope, the Poincaré map is obtained by intensifying the image on the screen at a certain phase of the forcing voltage (sometimes called "z-axis modulation"). In our laboratory, we were able to generate a 5-10 volt pulse of 1-2 µsec duration when the forcing function reached a certain phase:

$$\omega t_n = \phi_0 + 2\pi n \,. \tag{8}$$

This pulse was then used to intensify a phase plane image, $(x(t_n), \dot{x}(t_n))$, using two vertical amplifiers. Chaotic phase plane trajectories can often be unraveled using the Poincaré map by taking a set of pictures by varying ϕ_0 in (8) (see Figure 12). This is tantamount to sweeping the Poincaré plane in Figure 8. While one Poincaré map can be used to expose the fractal nature of the attractor, a complete set of maps varying ϕ_0 from 0 to 2π is sometimes needed to obtain a complete picture of the attractor on which the motion is riding.

One can also use a digital oscilloscope in an external sampling rate mode with the same narrow pulse signal used for the analog oscilloscope. A similar technique can be employed using an analog to digital (A-D) signal converter by storing the sampled data in a computer for display at a later time. The important point here is that the sampling trigger signal must be exactly synchronous with the forcing function.

2.3.1 Poincaré maps—effect of damping

If a system does not have sufficient damping, then the chaotic attractor will tend to uniformly fill up a section of phase space and the Cantor set structure which is characteristic of strange attractors will not be evident. An example of this is shown in Figure 10 for the vibration of a buckled beam. A comparison of low and high damping Poincaré maps shows that adding damping to the system can sometimes bring out the fractal structure.

On the other hand, if the damping is too high, the Cantor set sheets can appear to collapse onto one curve. In this case one can look for a one-dimensional map.

2.3.2 Construction of one-dimensional maps

There are a number of physical and numerical examples where the attracting set appears to have a sheet-like behavior in some three dimensional phase space. This often means that a Poincaré section, obtained by measuring the sequence of points which pierce a plane transverse to the attractor, will appear as a set of points along some one-dimensional line. This suggests that if one could parameterize these points along the line by a variable x, then it is possible a function exists which relates x_{n+1} and x_n, i.e.,

$$x_{n+1} = f(x_n) .$$

The function may be found by simply plotting x_{n+1} vs. x_n. The existence of such a function f(x) implies that the mathematical results for one-dimensional maps such as period doubling, Feigenbaum scaling, etc., may be applicable to the more complex physical problem in explaining, predicting or organizing experimental observations.

2.3.3 Position triggered Poincaré maps

However, in some problems, it is natural to trigger at a particular value of position as in an impact oscillator problem (see Figure 12). In such cases there are constraints on the position variable.

In the impact oscillator there are three convenient state variables, the position x, velocity v, and phase of the driving signal $\phi = \omega t$. If one triggers on the position at $x = x_0$, then the Poincaré map becomes a set of values $(v_n^{\pm}, \omega t_n)$ where $v_n^{\pm}$ is the velocity before or after impact and t_n is the time of impact. However, the results can be plotted in a cylindrical space where $0 < \omega t_n < 2\pi$.

An example of the experimental technique to obtain a (v_n, ϕ_n) Poincaré map is shown in Figure 13. When the mass hits the position constant, a sharp signal is obtained from a strain gauge or accelerometer. This sharp signal can be used to trigger a data storage device (such as a storage or digital oscilloscope) to store the value of the velocity of the particle.

To obtain the phase ϕ_n modulo 2π we generate a periodic ramp signal in phase with the driving signal where the minimum value of zero corresponds to $\phi = 0$, and the maximum voltage of the ramp corresponds to the phase $\phi = 2\pi$. The impact generated sharp spike voltage is used to trigger the data storage device and store the value of the ramp voltage along with the velocity signal before or after impact.

Another example of this kind of Poincaré map is shown in Figure 14 for the chaotic vibrations of a motor. In this problem the motor has a nonlinear torque-angle relation created by a d.c. current in one of the stator poles and the permanent magnet rotor is driven by a sinusoidal torque created by an a.c. current in an adjacent coil. The equation of motion for this problem is

$$J\ddot{\theta} + \tau\dot{\theta}_1 + \kappa \sin\theta = F_0 \cos\theta \cos\omega t . \tag{9}$$

To obtain a Poincaré map we choose a plane in the three-dimensional space $(\theta,\dot{\theta},\omega t)$ where $\theta = 0$ (Figure 15). This is done experimentally by using a slit in a thin disc attached to the rotor and using a light emitting diode and detector to generate a voltage pulse every time the rotor passes through $\theta = 0$. This pulse is then used to sample the velocity and measure the time. The data can be directly displayed on a storage oscilloscope or using a computer, can be replotted in polar coordinates as shown in Figure 16.

2.3.4 Double Poincaré maps

So far we have only talked of Poincaré maps for third order systems, such as a single degree of freedom with external forcing. But what about higher order systems with motion in a four or five dimensional phase space? For example, a two degree of freedom autonomous aeroelastic problem would have motion in a four-dimensional phase space (x_1,v_1,x_2,v_2), or if $x_1 \equiv x$, $(x(t_n), x(t_n - \tau), x(t_n - 2\tau), x(t_n - 3\tau))$. A Poincaré map triggered on one of the state variables would result in a set of points in a three-dimensional space. The fractal nature of this map if it exists might not be evident in three dimensions and certainly not if one projects this three-dimensional map onto a plane in two of the remaining variables.

A technique to observe the fractal nature of three-dimensional Poincaré map of a fourth order system has been developed in our laboratory which we call a *double Poincaré section* (see Moon and Holmes (1985)). (See Figure 17.) This technique enables one to slice a finite width section of the three-dimensional map in order to uncover fractal properties of the attractor and hence determine if it is "strange".

We illustrate this technique with an example derived from the forced motion of a buckled beam. In this case we examine a system with two incommensurate driving frequencies. The mathematical model has the form

$$\dot{x} = y$$
$$\dot{y} = -\tau y + F(x) + f_1 \cos \theta_1 + f_2 \cos(\theta_2 - \phi_0) \qquad (10)$$
$$\dot{\theta}_1 = \omega_1$$
$$\dot{\theta}_2 = \omega_2 .$$

The experimental apparatus for a double Poincaré section is shown in Figure 17. The driving signals were produced by identical signal generators and were added electronically. The resulting quasi-periodic signal was then sent to a power amplifier which drove the electromagnetic shaker.

The first Poincaré map was generated by a 1 μs trigger pulse synchronous with one of the harmonic signals. The Poincaré map (x_n, v_n) using one trigger results in a fuzzy picture with no structure as shown in Figure 18. To obtain the second Poincaré section, we trigger on the second phase of the driving signal. However, if the pulse width is too narrow, the probability of finding points coincident with the first trigger is very small. Thus we set the second pulse with 1000 times the first, at 1 ms. In the problem we looked at the second pulse width represents less than 1% of the second frequency phase of 2π. The (x,v) points were only stored when the first pulse was coincident with the second as shown in Figure 19. This was accomplished using a digital circuit with a logical NAND gate. Because of the infrequency of the simultaneity of both events, a map of 4000 points took upwards of ten hours compared to 8-10 minutes to obtain a conventional Poincaré map.

θ = 0°

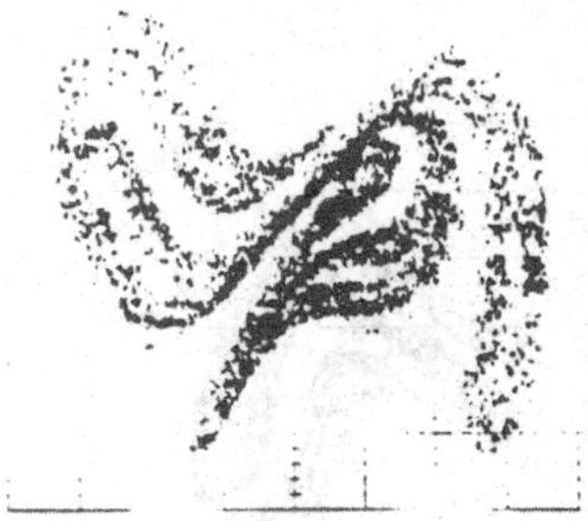

θ = 0°

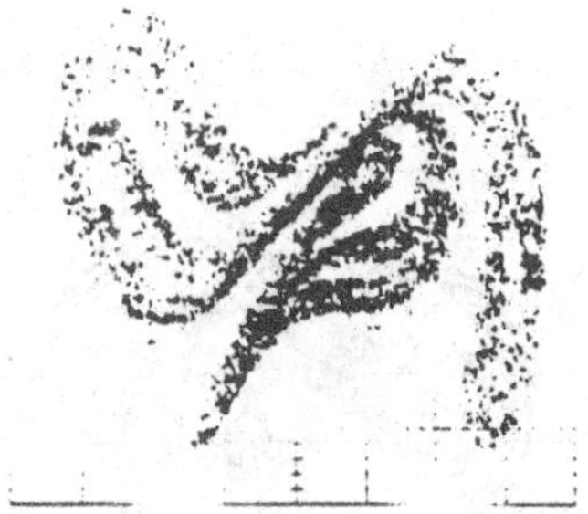

θ = 18°

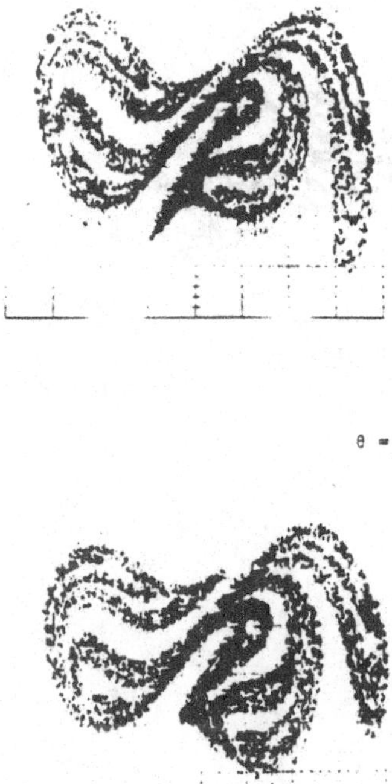

θ = 18°

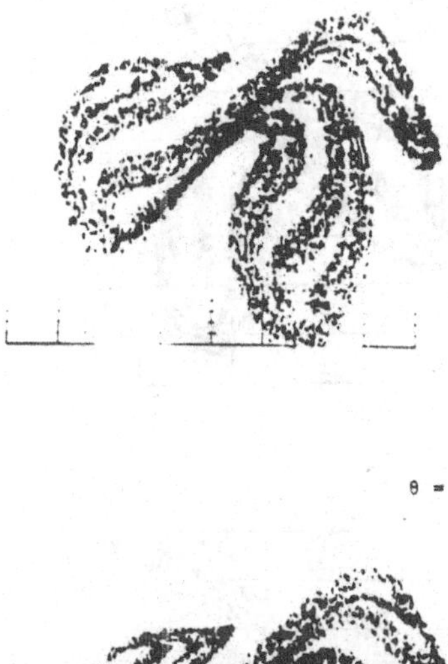

θ = 36°

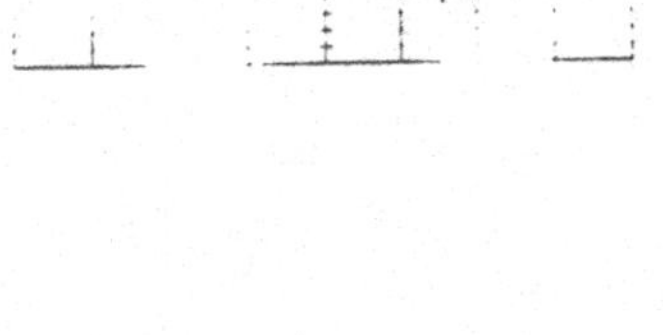

θ = 90°

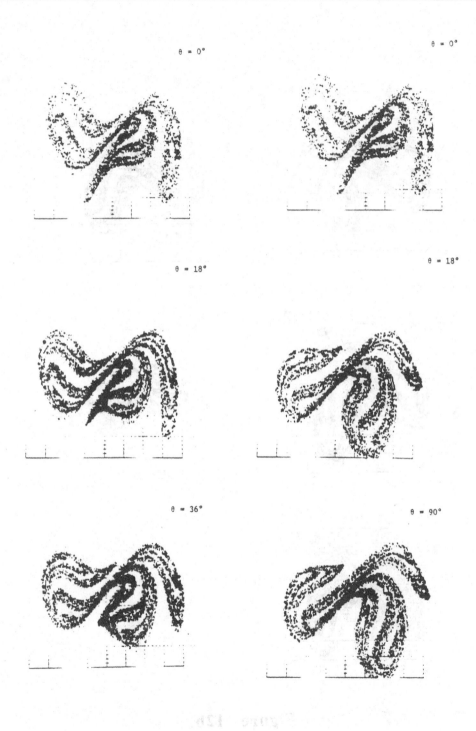

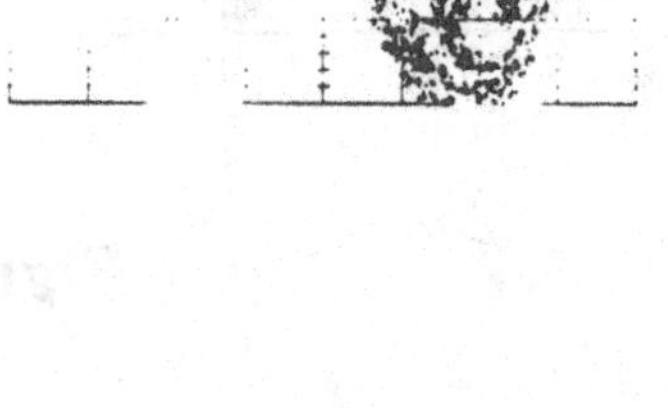

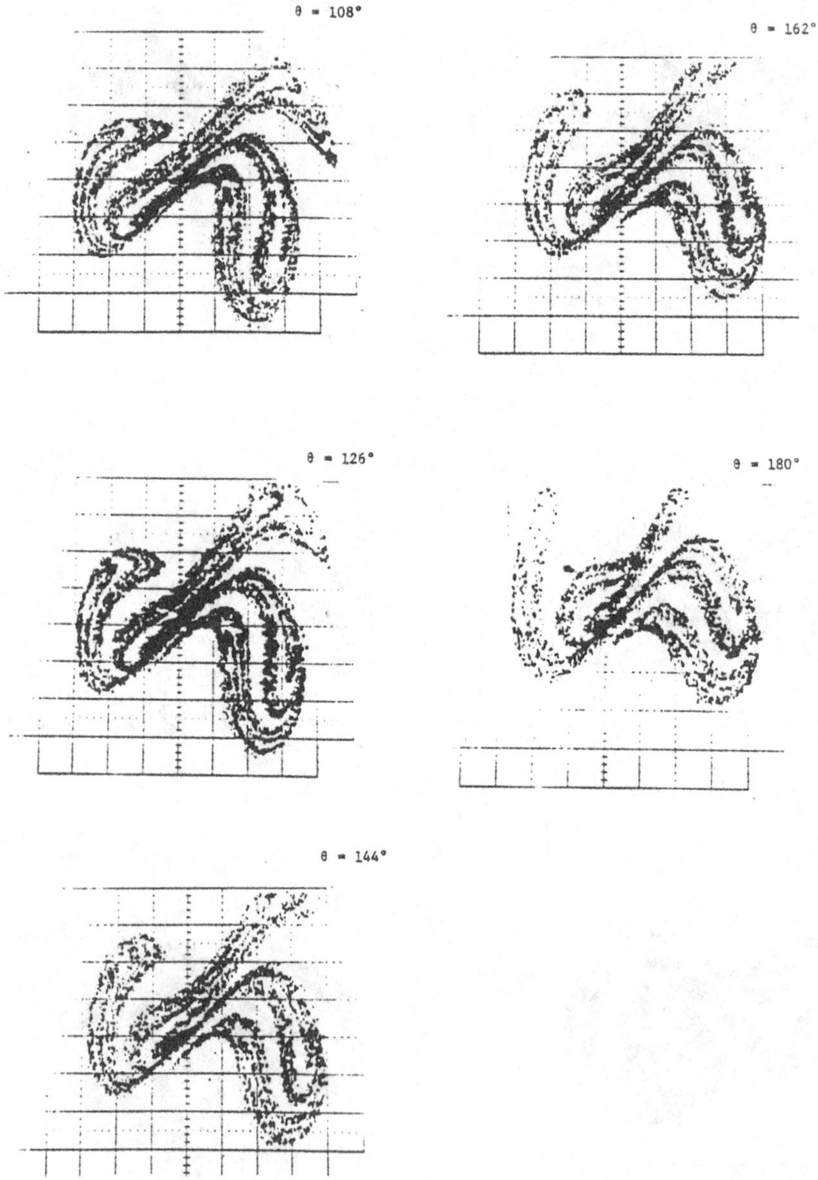

Figure 12b.

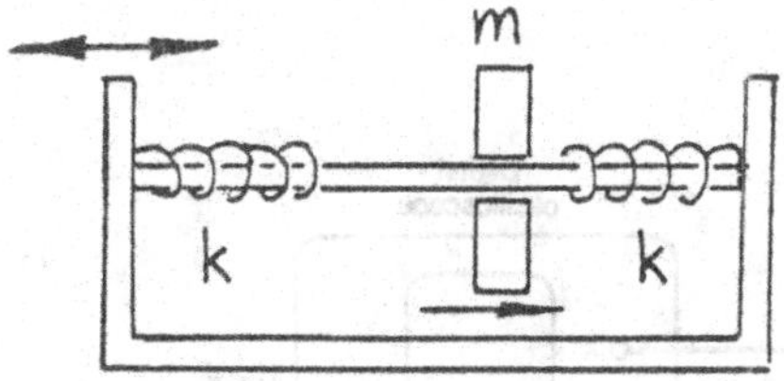

Figure 13a.

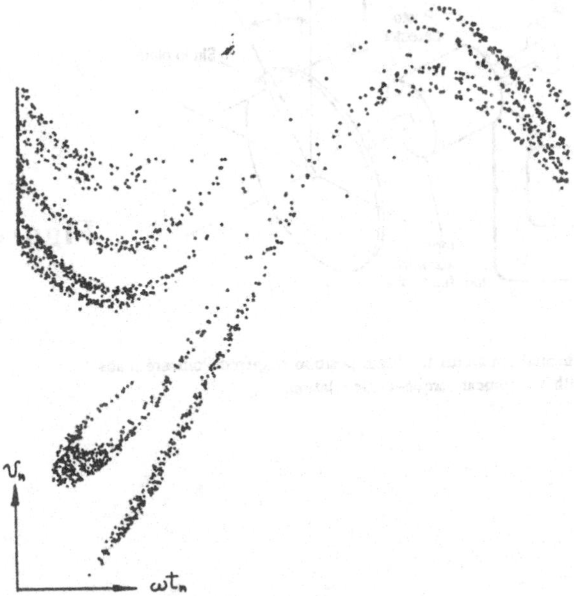

Position triggered Poincaré map for an oscillating mass with impact constraints

Figure 13b.

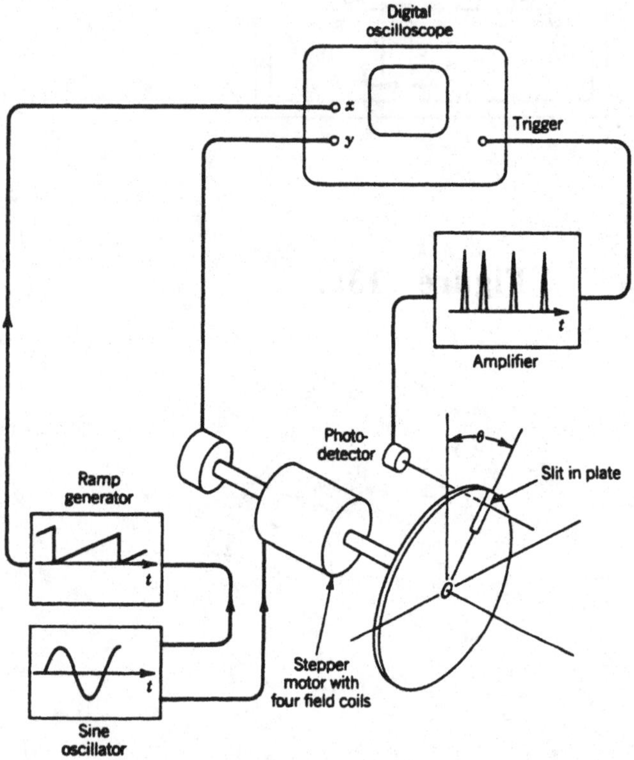

Figure 14.

Diagram of experimental apparatus to obtain position triggered Poincaré maps for a periodically forced rotor with a nonlinear torque–angle relation.

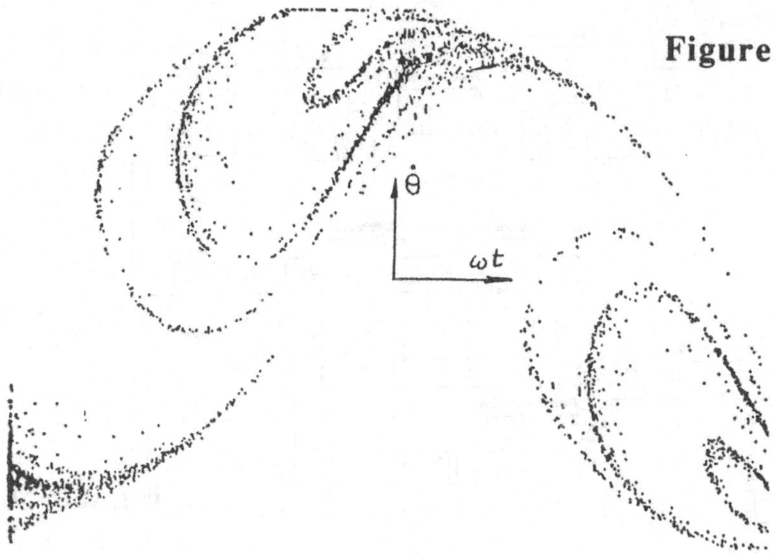

Figure 15.

Figure 16.

Position triggered Poincaré map for chaos in a nonlinear rotor (see Figure 14).

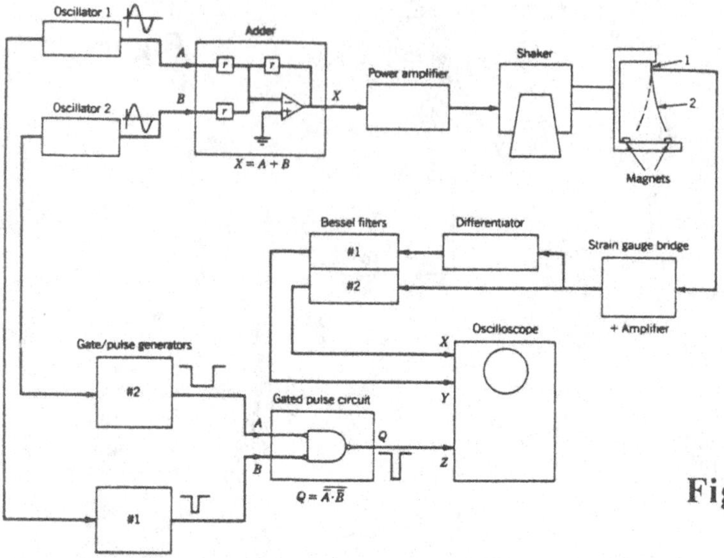

Figure 17.

Sketch of experimental apparatus to obtain Poincaré map for an oscillator with two driving frequencies [from Moon and Holmes (1985) with permission of Elsevier Science Publishers, copyright 1985]. Note: Strain gauges—1; steel beam—2.

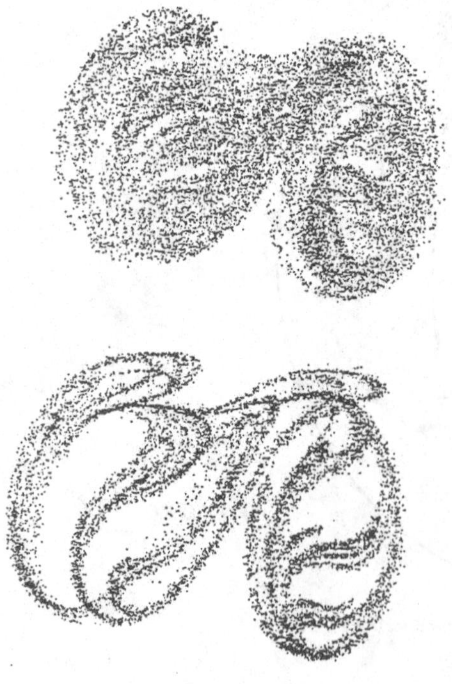

P_i^1 Map

$P^{(2)}$ Map

Figure 18.

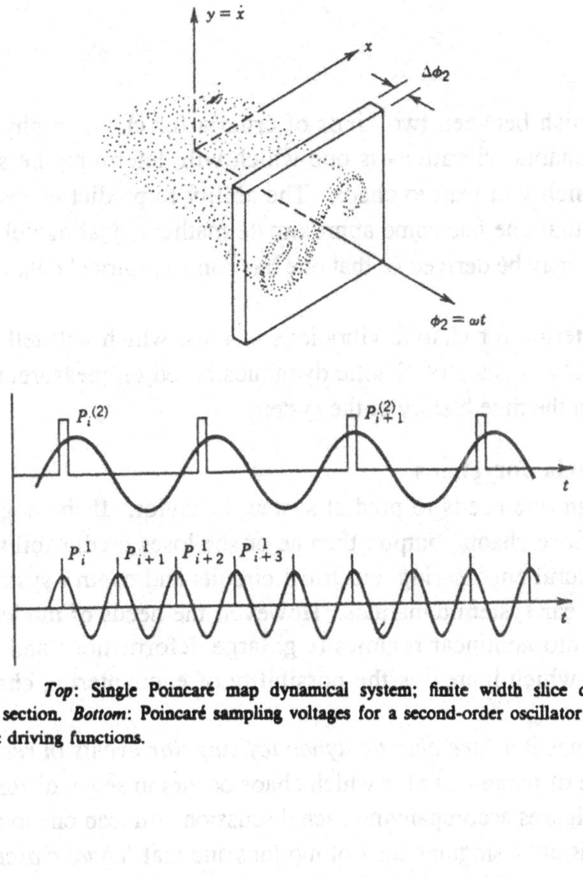

***Top*: Single Poincaré map dynamical system; finite width slice of second Poincaré section. *Bottom*: Poincaré sampling voltages for a second-order oscillator with two harmonic driving functions.**

Figure 19.

3. Chaos Diagrams

One can distinguish between two kinds of criteria for chaos in physical systems: a *predictive* rule for chaotic vibrations is one which will determine the set of input or control parameters which will lead to chaos. The ability to predict chaos in a physical system implies either that one has some approximate mathematical model of the system from which a criterion may be derived or that one has some empirical data based on many tests.

A *diagnostic* criterion for chaotic vibrations is a test which will tell if a particular system was or is in fact in a state of chaotic dynamics based on measurements or signal processing of data from the time history of the system.

3.1 Empirical criteria for chaos

In order to design one needs to predict system behavior. If the engineer chooses parameters which produce chaotic output, then he or she loses predictability. In the past many designs in structural engineering, electrical circuits and control systems were kept within the realm of linear system dynamics. However, the needs of modern technology have pushed devices into nonlinear regimes (e.g. large deformations and deflections in structural mechanics) which increases the possibility of encountering chaotic dynamic phenomena.

To address the question "*are chaotic dynamics singular events in real systems*" we will examine the range of parameters for which chaos occurs in seven different problems. A cursory scan of the figures accompanying each discussion will lead one to the conclusion that chaotic dynamics is not a singular class of motions and that "*chaotic oscillations occur in many nonlinear systems for a wide range of parameter values.*"

As an example consider the forced oscillations of a nonlinear inductor—Duffing's equation. The chaotic dynamics of a circuit with a nonlinear inductor have received extensive analog and digital simulation by Y. Ueda of Kyoto University (1980). The nondimensional equation, where x represents the flux in the inductor, takes the form

$$\ddot{x} + k\dot{x} + x^3 = B \cos t \qquad (11)$$

The time has been nondimensionalized by the forcing frequency so that the entire dynamics is determined by the two parameters k,B and the initial conditions ($x(0), \dot{x}(0)$). Here k is a measure of the resistance of the circuit, while B is a measure of the driving voltage. Ueda found that by varying these two parameters one could obtain a wide variety of periodic, subharmonic, ultrasubharmonic as well as chaotic motions. The regions of chaotic behavior in the (k,B) plane are plotted in Figure 20. The regions of subharmonic and harmonic motions are quite complex and only a few are shown for illustration. The Hatched areas with continuous lines indicate regions of chaos only, while the regions with broken lines indicate that both chaotic as well as periodic motion can occur depending on initial conditions. A theoretical criterion for this relatively simple equation has not been found to date.

Another example is the forced oscillations of a particle in a two-well potential—Duffing's equation. It was first studied by Holmes (1979) and later in a series of papers by the author and coworkers. The mathematical equation describes the forced motion of a particle between two states of equilibrium, which can be described by a two-well potential

$$\ddot{x} + \delta\dot{x} - \frac{1}{2}x(1 - x^2) = f \cos \omega t\,. \tag{12}$$

This equation can represent a particle in a plasma, defect in a solid and on a larger scale, the dynamics of buckled elastic beam. The dynamics are controlled by three nondimensional groups (δ,f,ω) where δ represents nondimensional damping, and ω is the driving frequency nondimensionalized by the small amplitude natural frequency of the system in one of the potential wells.

Regions of chaos from two studies are shown in Figures 21 and 22. The first represents experimental data for a buckled cantilevered beam. The ragged boundary is the experimental data while the smooth curve represents a theoretical criterion. The experimental criterion was determined by looking at Poincaré maps of the motion.

Results from numerical simulation of equation 12 are shown in Figure 22. The diagnostic tool used to determine if chaos was present was the Lyapunov exponent. This diagram shows that there are complex regions of chaotic vibrations in the plane (f,ω) for fixed damping δ. For very large forcing $f \gg 1$ one expects the behavior to emulate the previous problem studied by Ueda.

The theoretical boundary found by Holmes (1979) is discussed in the next section. It has special significance since below this boundary periodic motions are predictable while above this boundary one loses the ability to exactly predict which of many periodic or chaotic modes the motion will be attracted to. Above the theoretical criteria (based on homoclinic orbits) the motion is very sensitive to initial conditions, even when it is periodic.

A final example of a chaos diagram is the forced motions of a rotating dipole in magnetic fields—the pendulum equation. In this experiment a permanent magnet rotor is excited by crossed steady and time harmonic magnetic fields (see Moon, Cusamano, Holmes (1987)) as shown in Figure 23. The nondimensionalized equation of motion for the rotation angle θ resembles that for the pendulum in a gravitational potential:

$$\ddot{\theta} + \tau\dot{\theta} + \sin\theta = f\cos\theta\,\cos\omega t\,. \tag{13}$$

The regions of chaotic rotation in the f-ω plane, for fixed damping, are shown in Figure 23. This is one of the few published examples where both experimental and numerical simulation data are compared with a theoretical criterion for chaos. The theory is based on the homoclinic orbit criterion and is discussed in the next section. As in the case of the two-well potential, chaotic motions are to be found in the vicinity of the natural frequency for small oscillations (ω = 1.0 in Figure 23).

3.2 Lyapunov exponents

Thus far we have discussed mainly predictive criteria for chaos. Here we describe a tool for *diagnosing* whether or not a system is chaotic. Chaos in deterministic systems implies a sensitive dependence on initial conditions. This means that if two trajectories start close to one another in phase space, they will move exponentially away from each other for small times on the average. Thus if d_0 is a measure of the initial distance between the two starting points, at a later time the distance is

$$d(t) = d_0\, 2^{\lambda t}\,. \tag{14}$$

If the system is described by difference equations or a map, then we have

$$d_n = d_0\, 2^{\Lambda n}\,. \tag{15}$$

(The choice of base 2 in equations 14 and 15 is convenient but arbitrary). The symbols Λ,λ are called *Lyapunov exponents.*

An excellent review of Lyapunov exponents and their use in experiments to diagnose chaotic motion is given by Wolf, Swift et al. (1985). This review also contains two useful computer programs for calçulating Lyapunov exponents.

The divergence of chaotic orbits can only be locally exponential since if the system is bounded, as most physical experiments are, d(t) cannot go to infinity. Thus, to define a measure of this divergence of orbits we must average the exponential growth at many points along a trajectory as shown in Figure 24. One begins with a reference trajectory (called a *fiduciary* by Wolf et al. (1985)) and a point on a nearby trajectory and measures $d(t)/d_0$. When d(t) becomes too large (i.e. the growth departs from exponential behavior) one looks for a new "nearby" trajectory and defines a new $d_0(t)$. Once can define the first Lyupunov exponent by the expression

$$\lambda = \frac{1}{t_N - t_0} \sum_{k=1}^{N} \log_2 \frac{d(t_K)}{d_0(t_K - 1)} . \tag{16}$$

Then the criterion for chaos becomes

$$\begin{aligned} &\lambda > 0 \ \text{chaotic} \\ &\lambda \leq 0 \ \text{regular motion} . \end{aligned} \tag{17}$$

The reader by now has surmised that this operation can only be done with the aid of a computer whether the data is from a numerical simulation or from a physical experiment.

Only in a few pedogogical examples can one calculate λ explicitly. To examine one such case consider the extension of the concept of Lyapunov exponents to a one-dimensional map.

$$x_{n+1} = f(x_n) . \tag{18}$$

In regions where f(x) is smooth and differentiable, the stretch between neighboring orbits is measured by $|df/dx|$. To see this suppose we consider two initial conditions x_0, $x_0+\varepsilon$. Then in (15)

$$\begin{aligned} &d_0 = \varepsilon \\ &d_1 = f(x_0 + \varepsilon) - f(x_0) \approx \left.\frac{df}{dx}\right|_{x_0} \varepsilon . \end{aligned} \tag{19}$$

Following (16) we define the Lyapunov or characteristic exponent as

$$\Lambda = \lim_{N \to \infty} \frac{1}{N} \sum_{k=0}^{N} \log_2 \left| \frac{df}{dx}(x_n) \right| . \tag{20}$$

An illustrative example is the *Bernoulli map*

$$x_{n+1} = 2x_n \qquad \text{(modulus 1)} . \tag{21}$$

Here (mod 1) means

$$x(\text{mod } 1) = x - \text{Integer}(x) .$$

This map is multivalued and is known to be chaotic. Except for the switching value at $x = 1/2$, $|f'| = 2$. Applying the definition (20) we find $\Lambda = 1$. Thus on the average the distance between nearby points grows as

$$d_n = d_0 2^n .$$

The units of Λ are one bit per iteration. One interpretation of Λ is that one bit of information about the initial state is lost every time the map is iterated. To see this write x_n in binary notation. For example $x_n = \left(\frac{1}{2} + \frac{1}{16} + \frac{1}{128}\right) \equiv 0.1101001$ and x(mod 1) means 1.101001 (mod 1) = 0.101001. Thus the map $2x_n$ (mod 1) moves the decimal point to the right and drops the integer value. So if we start out with m significant decimal places of information we lose one each iteration, i.e., we lose one bit of information. *After m iterations we have lost knowledge of the initial state of the system.*

It can be shown that the solution for the logistic or quadratic map becomes chaotic when the control parameter $a > 3.57$:

$$x_{n+1} = ax_n(1 - x_n) . \qquad (22)$$

This can be verified by calculating the Lyapunov exponent as a function of "a" as shown in Figure 25. Beyond a = 3.57 the exponent becomes nonpositive in the periodic windows between $3.57 < a < 4$. When a = 4 it has been shown that $\lambda = \ln 2$ (see e.g. Schuster (1985)).

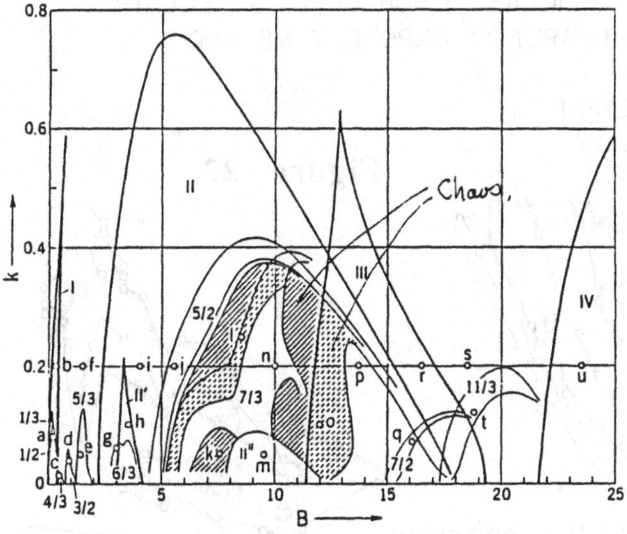

steady states for the system exhibited by Eq. (1).

Figure 20.

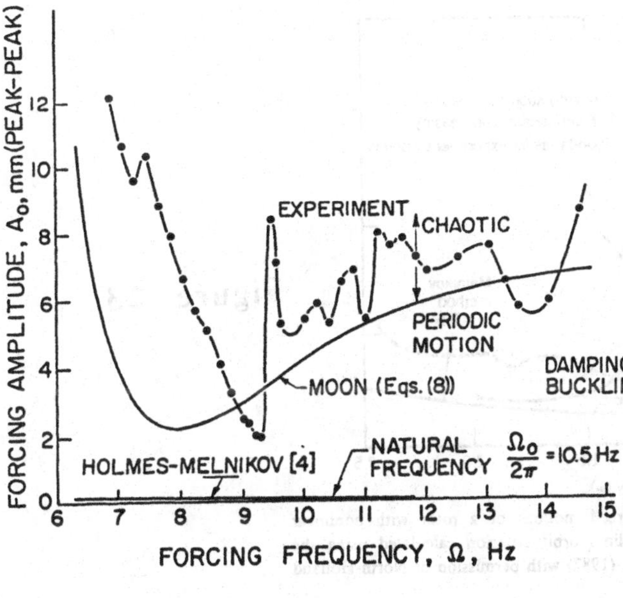

Figure 21.

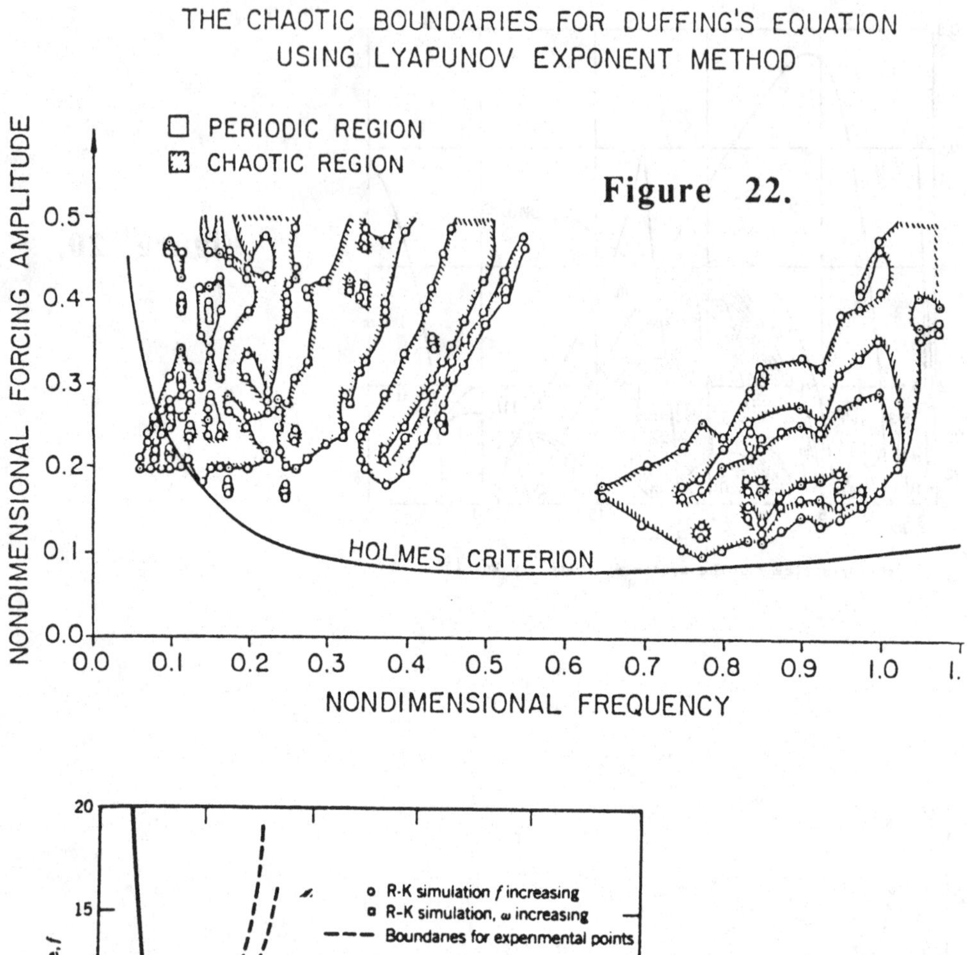

Figure 22.

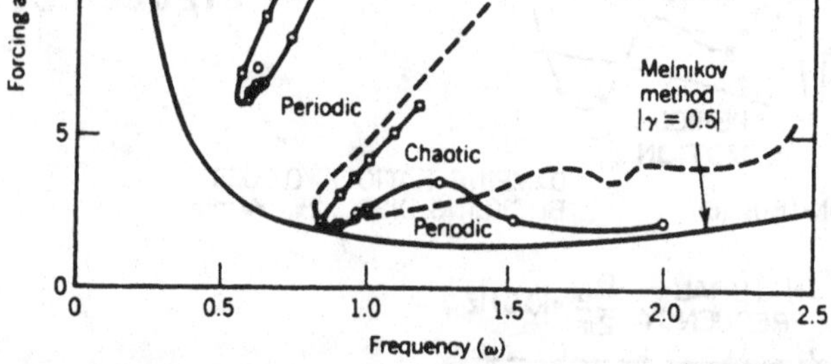

Figure 23.

Experimental chaos diagram for forced motions of a rotor with nonlinear torque–angle property. Comparison with homoclinic orbit criterion calculated using the Melnikov method (Section 5.3) [From Moon et al. (1987) with permission of North-Holland Publishing Co., copyright 1987].

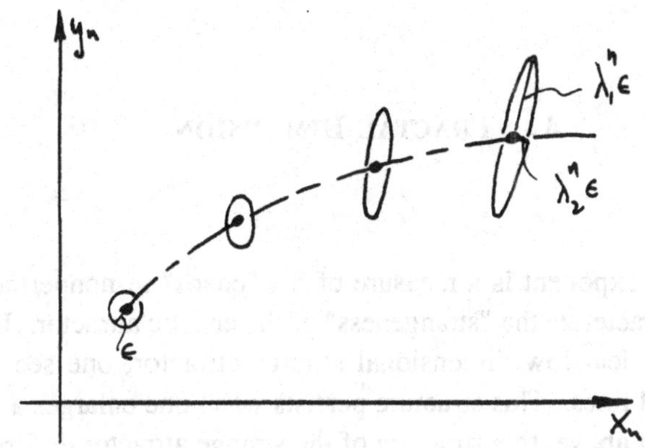

Figure 24.

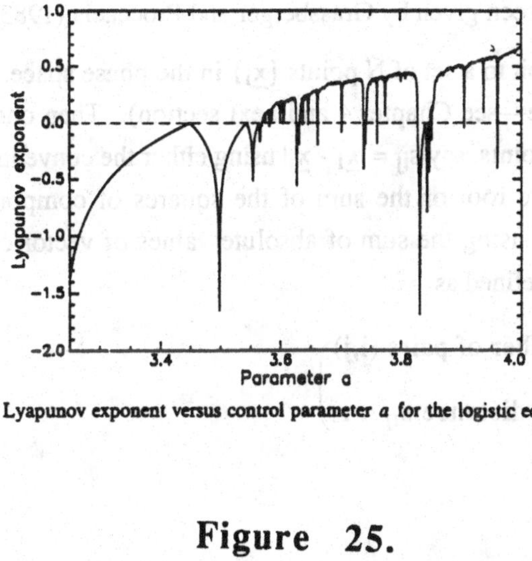

Lyapunov exponent versus control parameter *a* for the logistic equation (5-4.9).

Figure 25.

4. Fractal Dimension

If the Lyapunov exponent is a measure of the "chaos" in nonperiodic motion, we would also like to characterize the "strangeness" of the chaotic attractor. If one looks at a Poincaré map of a typical low dimensional strange attractor, one sees sets of points arranged along parallel lines. This structure persists when one enlarges a small region of the attractor. As noted above, this structure of the strange attractor differs from periodic motions (just a finite set of Poincaré points) or quasi-periodic motion which in the Poincaré map becomes a closed curve. In the Poincaré map one can say that the dimension of the periodic map is zero and the dimension of the quasi-periodic map is one. The idea of the fractal dimension calculation is to attach a measure to the Cantor-like set of points in the strange attractor. If the points uniformly covered some area on the plane as in Figure 10a, we might say its dimension was close to two. Because the chaotic map in Figure 10b has an infinite set of gaps, its dimension is between one and two—thus the word *fractal* dimension.

4.1 Correlation dimension

This measure of fractal dimension has been successfully used by experimentalists (see e.g., Malraison et al. (1983) and Moon and Li (1985)). An extensive study of this definition of dimension has been given by Grassberger and Proccacia (1983).

One discretizes the orbit to a set of N points $\{\underset{\sim}{x}_i\}$ in the phase space. (One can also create a pseudo-phase space—see Chapter 4 and next section). Then one calculates the distances between pairs of points, say $s_{ij} = |\underset{\sim}{x}_i - \underset{\sim}{x}_j|$ using either the conventional Euclidean measure of distance (square root of the sum of the squares of components) or some equivalent measure such as using the sum of absolute values of vector components. A correlation function is then defined as

$$C(r) = \lim_{N \to \infty} \frac{1}{N^2} \begin{pmatrix} \text{Number of pairs } (i,j) \\ \text{with distance } s_{ij} < r \end{pmatrix} . \qquad (23)$$

For many attractors this function has been found to exhibit a power law dependence on r as $r \to 0$, i.e.,

$$\lim_{r \to 0} C(r) = ar^d$$

so that one may define a fractal or correlation dimension using the slope of the ln C vs. ln r curve

$$d = \lim_{r \to 0} \frac{\ln C(r)}{\ln r}. \tag{24}$$

It has been shown that C(r) may be calculated more effectively by constructing a sphere or cube at each point $\underset{\sim}{x}_i$ in phase space and counting the number of points in each sphere, i.e.,

$$C(r) = \lim_{r \to 0} \frac{1}{N^2} \sum_{i}^{N} \sum_{\substack{j \\ i \neq j}}^{N} H(r - |\underset{\sim}{x}_i - \underset{\sim}{x}_j|) \tag{25}$$

where $H(s) = 1$ if $s > 0$ and $H(s) = 0$ if $s < 0$. m One can think of the first sum as a probability measure P. A variation of this method is to average $P(r;x_i) \equiv P_i$ over a smaller set of points than N, so that

$$d = \lim_{r \to 0} \frac{\ln \frac{1}{m} \Sigma P_i}{\ln r}.$$

In using this and other measures of dimension, one must be careful not to choose the radius of the sphere either too small or too large. One can use the measure of distance r about each point in the form $|\underset{\sim}{s}_i| = |s_x| + |s_y|$ so that the "sphere" about each point is effectively a diamond-shaped area within which the number of points are counted.

4.2 Fractal dimension of strange attractors

There are two principal applications of fractal mathematics to nonlinear dynamics: characterization of strange attractors and measurement of fractal boundaries in initial condition and parameter space. In this section we discuss the use of the fractal dimension in both numerical and experimental measurements of motions associated with strange attractors.

As yet there are no instruments, electronic or otherwise, which will produce an output proportional to the fractal dimension, although electro-optical methods may achieve this end in the future. To date, in both numerical and experimental measurements, the fractal dimension as well as Lyapunov exponents are found by discretizing the signals at uniform time intervals and the data is processed with a computer. There are three basic methods:

i) time discretization of phase space variables
ii) calculation of fractal dimension of Poincaré maps
iii) construction of pseudo-phase space based on discretization of single variable measurements

In both the first and third methods the variables are measured and stored at uniform time intervals $\{\underset{\sim}{x}_i (t_0 + n\tau)\}$, where n is a set of integers. If the Poincaré map in ii) is based on a time signal, then the τ is just the period of the time based Poincaré map. However, if the Poincaré map is based on other phase space variables, then the data are collected at variable times depending on the specific type of Poincaré map.

There are three principal definitions of fractal dimension used today: averaged pointwise dimension, correlation dimension and Lyapunov dimension. In most of the current experience with actual calculation of fractal dimension, between 2,000 and 20,000 points are used, though several recent papers claim to have reliable algorithms based on as little as 300 points. Direct algorithms for calculating fractal dimension based on N_0 points generally take N_0^2 operations so that supermini- or mainframe computers are often used. However, clever use of basic machine operations can reduce the number of operations to order $N_0 \ln N_0$ and significantly speed up calculation.

i) Discretization of Phase Space Variables

Suppose we know or suspect a chaotic system to have an attractor in three-dimensional phase space based on the physical variables $\{x(t), y(t), z(t)\}$. For example, in the case of the forced motion of a beam or particle in a two-well potential x = position, $v = \dot{x}$ is the velocity, and $z = \omega t$ is the phase of the periodic driving force. In this method time samples of (x(t), y(t), z(t)) are obtained at a rate that is smaller than the driving force period. To each time interval corresponds a point $\underset{\sim}{x}_n = (x(n\tau), y(n\tau), z(n\tau))$ in phase space.

To calculate an averaged pointwise dimension, one chooses a number of random points $\underset{\sim}{x}_n$. About each point one calculates the distances from $\underset{\sim}{x}_n$ to the nearest points surrounding $\underset{\sim}{x}_n$. (Note these points are not the nearest in time, but in distance). One does not need to use a Euclidean measure of distance. For example, the sum of absolute values of the components of $(\underset{\sim}{x}_n - \underset{\sim}{x}_m)$ could be used, i.e.,

$$s_{im} = |x(i\tau) - x(m\tau)| + |y(i\tau) - y(m\tau)| + |z(i\tau) - z(m\tau)| . \qquad (26)$$

Then the number of points within a ball, cube or other geometric shape of order ε is counted and a probability measure is found as a function of ε

$$P_i(\varepsilon) = \frac{1}{N_0} \sum_{m=1} H(\varepsilon - s_{im}) \qquad (27)$$

where N_0 is the total number of sampled points and H is the Heaviside step function: $H(r) = 1$ if $r > 0$: $H(r) = 0$ if $r < 0$. The averaged pointwise dimension, following (27), is then

$$d_n = \lim_{r \to 0} \frac{\ln P_n(\varepsilon)}{\ln \varepsilon}$$

$$d \equiv \frac{1}{M} \sum_{n+1}^{M} d_n \qquad (28)$$

where the limit defining d_n exists. For some attractors, the function P_n vs. ε is not a power law but has steps or abrupt changes in slope. Then one can calculate a modified correlation dimension by first averaging P_n. For example, let

$$\hat{C}(\varepsilon) = \frac{1}{M} \sum^{M} P_n(\varepsilon)$$

$$d_n = \lim_{\varepsilon \to 0} \frac{\ln \hat{C}(\varepsilon)}{\ln \varepsilon} . \qquad (29)$$

This is similar to the correlation dimension discussed in the previous section.

In practice if $N_0 \sim 2{,}000 - 20{,}000$ points, then the number of sampled pointwise dimensions should be around $0.20\ N_0$. One should, of course, start with a small value of M and increase it to see if the average d reaches some limit.

The choice of ε also requires some judgment. The upper limit of ε is much smaller than the maximum size of the attractor yet large enough to capture the larger scale structure in the vicinity of the point x_n. The smallest value of ε must be such that the associated sphere or cube contains at least one sample point.

Another constraint on the minimum size of ε is the "real noise" or uncertainty in the measurements of the state variables (x,y,z). In an actual experiment there is a sphere of uncertainty surrounding each measured point in phase space. When ε becomes smaller than the radius of this sphere, the theory of fractal dimension discussed above comes into question since for smaller ε one cannot expect a self-similar structure.

ii) Fractal dimension of Poincaré maps

In systems driven by a periodic excitation, as in the Duffing-Ueda strange attractor (11-) or the two-well potential strange attractor (12-), time or the phase $\phi = \omega t$ becomes a natural phase space variable. In most cases, this time variable will lie in the attractor subspace and time can be considered as one of the contributions to the dimension of the attractor. In the case of a periodically forced nonlinear second order oscillator, the Poincaré map based on periodic time samples produces a distribution of points in the plane. To calculate the fractal dimension of the complete attractor, it is sometimes convenient to calculate the fractal dimension of the Poincaré map $0 < D < 2$. If D is independent of the phase of the Poincaré map (remember $0 \leq \omega t \leq 2\pi$), then the dimension of the complete attractor is just

$$d = 1 + D \,. \tag{30}$$

As an example we present numerical and experimental data for the two-well potential or Duffing-Holmes strange attractor

$$\ddot{x} + \tau \dot{x} - \frac{1}{2} x(1 - x^2) = f \cos \omega t) \,. \tag{31}$$

In this example we are interested in two questions:

a) Does the fractal dimension of the strange attractor vary with the phase of the Poincaré map?

b) How does the fractal dimension vary with the damping τ?

An experimental Poincaré map for the case of the buckled beam under periodic excitation is shown in Figure 26. The correlation function $C(\varepsilon)$ vs ε is shown plotted in a log-log scale and shows a linear dependence as assumed in the theory. The fractal dimension was calculated for a set of Poincaré maps and shows an almost constant value around the attractor. Thus the assumption $d = 1 + D$ in (30) appears to be a good one.

The effect of damping on the fractal dimension of the two-well potential strange attractor was determined from Runge-Kutta numerical simulation (see Moon and Li (1985)). This dependence is shown in Figure 27. The data show that low damping yields an attractor that fills phase space ($D = 2$, $d = 3$) as would a Hamiltonian (zero damping) system. As damping is increased, however, the Poincaré map looks one dimensional and the attractor has a dimension close to $d = 2$ as in the case of the Lorenz equations.

The fractal dimension of a chaotic circuit (diode, inductor and resistor in series driven with an oscillator) has beren measured by Lindsay (1985) using a Poincaré map. He measures the current at a sampling time equal to the period of the oscillator and constructs a three dimensional pseudo-phase space using $(I(t), I(t+\tau), I(t+2\tau))$ (see next section). He obtains a fractal dimension of the Poincaré map of $D = 1.58$ and infers a dimension of the attractor of 2.58.

iii) Dimension calculation from time series measurement

The methods discussed above assume that a) one knows the dimension of the phase space wherein the attractor lies, and b) that one has the ability to measure all the state variables. However, in many experiments the time history of only one state variable may be available or possible. Also in continuous systems involving fluid or solid continua, the number of degrees of freedom or minimum number of significant modes contributing to the chaotic dynamics may not be known a priori. In fact, one of the important applications of fractal mathematics is to allow one to determine the smallest number of first order differential equations that may capture the qualitative features of the dynamics of continuous systems. This has already had some success in thermo-fluid problems such as Rayleigh-Bernard convection (see Malraison et al. (1983)).

In early theories of turbulence, e.g., Landau, 1941, it was thought that chaotic flow was the result of the interaction of a very large or infinite set of modes or degrees of freedom in the fluid. At the present time, it is believed that the chaos associated with the transition to turbulence can be modelled by a finite set of ordinary differential equations.

Thus suppose that the number of first order equations required to simulate the dynamics of a dissipative system is N_0. Then the fractal dimension of the attractor would be $d < N_0$. Then if we were to determine d by some means, we would then determine the minimum N_0.

Not knowing N_0, we cannot know how many physical variables $(x(t), y(t), z(t), ...)$ to measure. Instead we construct a pseudo-phase space (embedding space) using time delayed measurements of one physical variable, say $(x(t), x(t+\tau), x(t+2\tau), ...)$. For example, three-dimensional pseudo-phase space vectors are calculated using three successive components of the digitized x(t), i.e.,

$$\underset{\sim}{x}_n = \{x(t_0 + n\tau), x(t_0 + (n+1)\tau), x(t_0 + (n+2)\tau)\} . \tag{32}$$

Using these position vectors, one can use the correlation function or averaged probability function to calculate a fractal dimension.

To determine the minimum N, one constructs higher dimensional pseudo-phase spaces based on the time sampled x(t) measurements until the value of the fractal dimension reaches an asymptote, say, $d = M + \mu$ where $\mu < 1$. Then the minimum phase space dimension for this chaotic attractor is $N = M+1$.

According to the theory of Takens (1978) if the dimension of the phase space of the original attractor is N, then one has to calculate d in an embedding space of dimension 2N+1.

As an example we describe some unpublished work by Cusumano and Moon from Cornell University. The physical system involves the forced vibration of a flexible cantilevered beam (known as the elastica) (Figure 28). For large enough motion, the beam vibrations will jump out of the planar mode into a nonlinear, nonplanar mode that appears chaotic. Using 20,000 time sampled measurements of the strain at the damped end, we used the embedding space method and the modified correlation dimension to measure the fractal dimension. The data in Figure 29 shows that this dimension is between 4 and 5. This suggests that the complex motion may be described by a two-degree-of-freedom nonlinear oscillator model.

Fleur de Poincaré

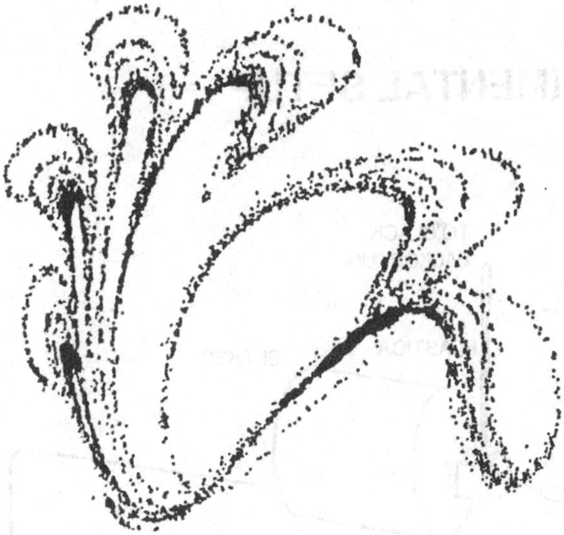

Figure 26.

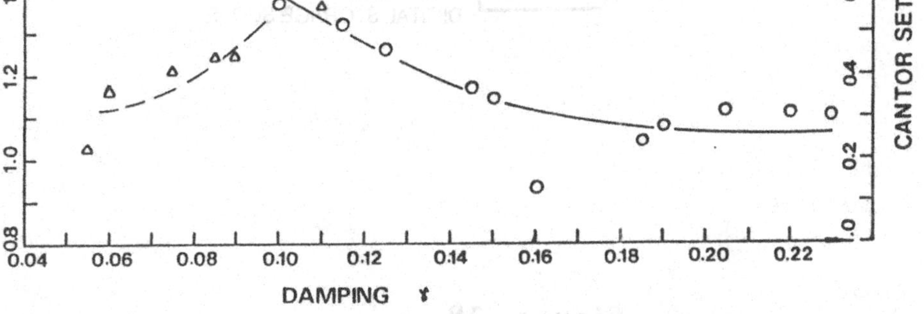

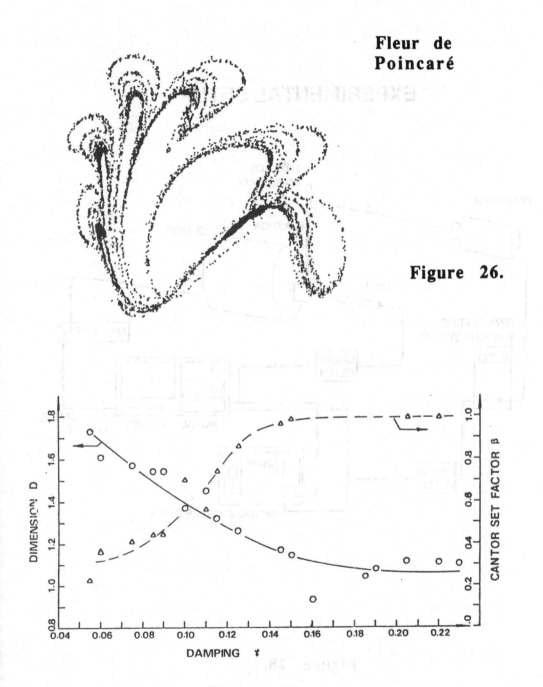

Figure 27.

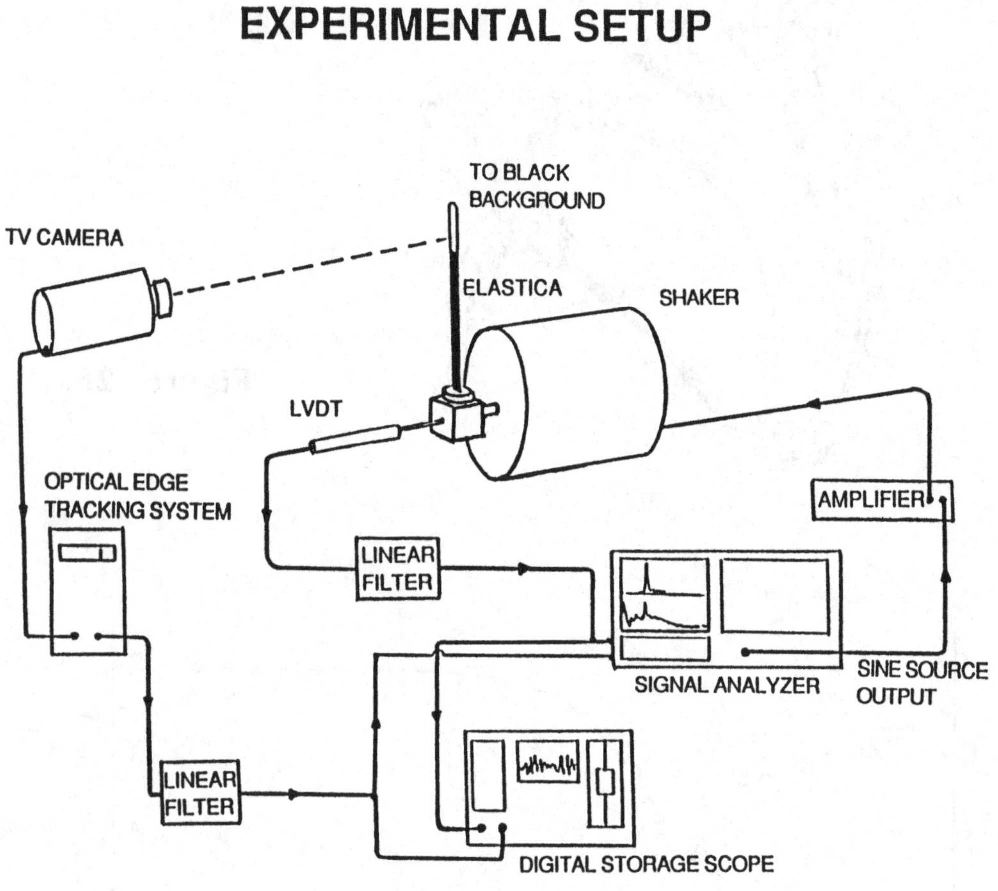
EXPERIMENTAL SETUP
TO BLACK
BACKGROUND
TV CAMERA
ELASTICA
SHAKER
LVDT
OPTICAL EDGE
TRACKING SYSTEM
AMPLIFIER
LINEAR
FILTER
SIGNAL ANALYZER
SINE SOURCE
OUTPUT
LINEAR
FILTER
DIGITAL STORAGE SCOPE

Figure 28.

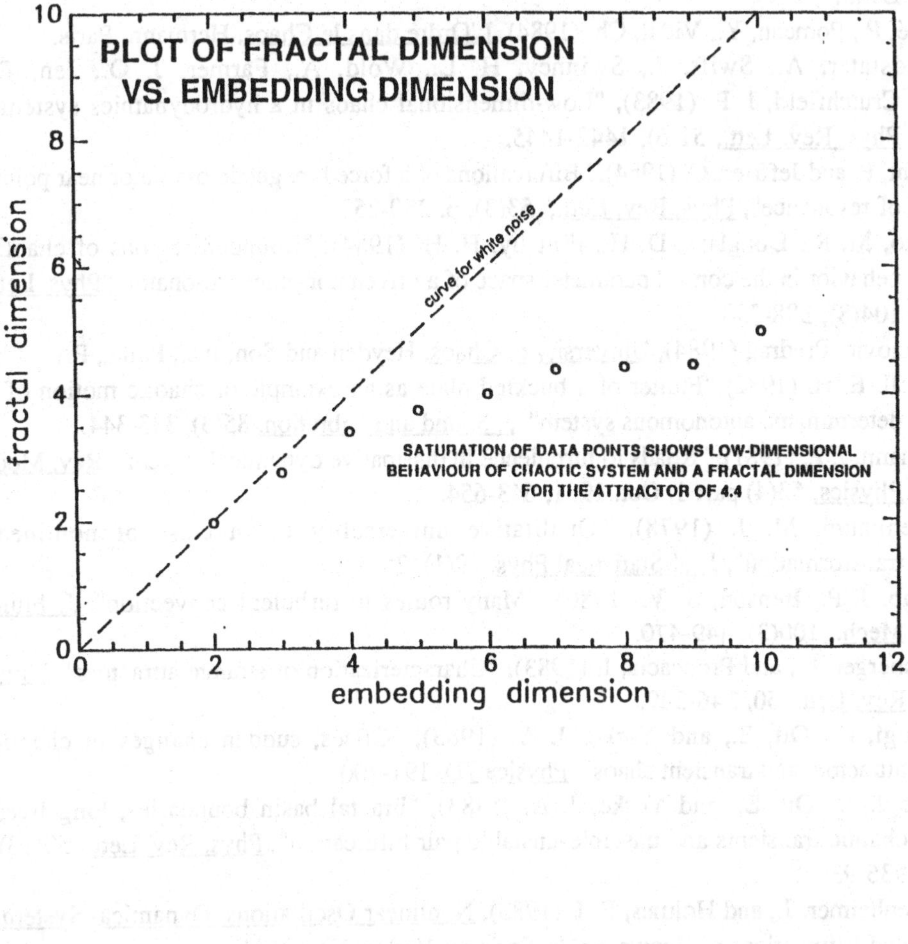
PLOT OF FRACTAL DIMENSION
VS. EMBEDDING DIMENSION
curve for white noise
SATURATION OF DATA POINTS SHOWS LOW-DIMENSIONAL
BEHAVIOUR OF CHAOTIC SYSTEM AND A FRACTAL DIMENSION
FOR THE ATTRACTOR OF 4.4
fractal dimension
embedding dimension
0
2
4
6
8
10
12

Figure 29.

REFERENCES

Bergé, P., Dubois, M., Manneville, P., and Pomeau, P. (1980), "Intermittancy in Rayleigh-Benard convection", Le J. de Physique-Letters, 41(15) 1, Aug., L341-L345.

Bergé, P., Pomeau, Y., Vidal, Ch. (1984), L'Ordre dans le Chaos, Hermann, Paris.

Brandstater, A., Swift, J., Swinney, H. L., Wold, A., Farmer, J. O., Jen, E., Crutchfield, J. P. (1983), "Low-dimensional chaos in a hydrodynamics system", Phys. Rev. Lett., 51(6), 1442-1445.

Bryant, P. and Jeffries, C. (1984), "Bifurcations of a forced magnetic oscillator near points of resonance", Phys. Rev. Lett., 53(3), p. 250-253.

Bucko, M. R., Douglass, D. H., Frutchy, H. H. (1984), "Bounded regions of chaotic behavior in the control parameter space of a driven non-linear resonator", Phys. Lett. 104(8), 388-390.

Cvitanovic, Predrag (1984), University in Chaos, Heyden and Son, Inc., Phila., PA.

Dowell, E. H. (1982) "Flutter of a buckled plate as an example of chaotic motion of a deterministic autonomous system", J. Sound and Vibration, 85(3), 333-344.

Eckmann, J. P. (1981), "Roads to turbulence in dissipative dynamical system", Rev. Mod. Physics, 53(4) part 1, Oct. 1981, 643-654.

Feigenbaum, M. J. (1978), "Qualitative universality for a class of nonlinear transformation", J. of Statistical Phys. 19(1), 25-52.

Gollub, J. P., Benson, S. V. (1980), "Many routes to turbulent convection", J. Fluid Mech., 100(3), 449-470.

Grassberger, P., and Proccacia, I. (1983), "Characterization of strange attractors", Phys. Rev. Lett., 50, 346-349.

Grebogi, C., Ott, E., and Yorke, J. A. (1983), "Crises, sudden changes in chaotic attractors and transient chaos", Physica 7D, 191-200.

Grebogi, C., Ott, E., and Yorke, J. A. (1983), "Fractal basin boundaries, long lived chaotic transients and unstable-unstable pair bifurcation", Phys. Rev. Lett., 50(13), 935-938.

Guckenheimer, J., and Holmes, P. J. (1983), Nonlinear Oscillations, Dynamical Systems and Bifurcations of Vector Fields, Springer Verlag, New York.

Hao, B.-L. (1984), Chaos, World Scientific Publ., Singapore.

Helleman, R. H. G. (1980), "Self-generated chaotic behavior in nonlinear mechanics", in Fundamental Problems in Statistical Mechanics, 5, 165-233.

Hendriks, F. (1983), "Bounce and chaotic motion in print hammers", IBM J. of Research and Development, 27(3), 273-280.

Henon, M. (1976), "A two-dimensional map with a strange attractor", Commun. Math. Phys., 50, 69.

Holmes, P. J. (1981), Editor, New Approaches to Nonlinear Problems in Dynamics, published by SIAM, Phila., PA.

Holmes, P. (1984), "Bifurcation sequences in horseshoe maps: infinitely many routes to chaos", Phys. Lett., 104A(6,7), 299-302.

Holmes, P. J., Moon, F. D. (1983), "Strange attractors and chaos in nonlinear mechanics", J. Applied Mechanics 50, 1021-1032.

Holmes, P. J. (1979), "A nonlinear oscillator with a strange attractor," Phil. Trans. Roy. Soc. A, 292, pp. 419-448.

Holmes, P. J. (1982), "The dynamics of repeated impacts with a sinusoidally vibrating table," J. Sound Fib., 84, pp. 173-189.

Iooss, G., and Joseph, D. D. (1980), Elementary Stability and Bifurcation Theory, Springer-Verlag, New York.

Kadanoff, L. P. (1983), "Roads to chaos", Physics Today (Dec.), 46-53.

Lee, C.-K., Moon, F. C. (1986), "An optical technique for measuring fractal dimensions of planar Poincaré maps", Phys. Lett., 114A(5), 222-226.

Leven, P. W., and Koch, B. P. (1981), "Chaotic behavior of a parametrically excited damped pendulum", Phys. Lett. 86A(2), 2 Nov., 71-74.

Lichtenberg, A. J., and Lieberman, M. A. (1983), Regular and Stochastic Motion, Springer-Verlag, New York.

Linsay, P. S. (1981), Period doubling and chaotic behavior in a driven, anharmonic oscillator," Phys. Rev. Lett., 47, No. 19, 1349-1352.

Lorenz, E. N. (1963), "Deterministic non-periodic flow", J. Atmospheric Sciences, 20, pp. 130-141.

Malraison, G., Atten, P., Bergé, P., and Dubois, M. (1983), "Dimension of strange attractors: an experimental determination of the chaotic regime of two convective systems", J. Physique-Lettres 44, 897-902.

Mandelbrot, B. B. (1977), Fractals, W. H. Freeman & Co., San Francisco, CA.

Manneville, P., Pomeau, Y. (1980), "Different ways to turbulence in dissipative dynamical systems", Physica 1D, 219-226.

Matsumoto, T., Chua, L. O., Komuro, M (1985), "The double scroll", IEEE Trans. Circuits and Systems, CAS-32, No. 8, August 1985, 798-818.

McDonald, S. W., Grebogi, C., Ott, E., and Yorke, J. A. (1985), "Fractal basin boundaries", Physica 17D, 125-153.

Miles, J. (1984), "Resonant motion of spherical pendulum", Physica 11D, 309-323.

Minorsky, N. (1962), Nonlinear Oscillations, D. Van Nostrand Co. Inc., Princeton, NJ.

Moon, F. C., Cusumano, J., Holmes, P. J. (1986), "Evidence for homoclinic orbits as a precursor to chaos in a magnetic pendulum", Physica D.
Moon, F. C. (1987), Chaotic Vibrations, J. Wiley and Sons, NY.
Moon, F. C., Holmes, W. T. (1985), "Double Poincaré sections of a quasi-periodically forced, chaotic attractor", Phys. Lett., 111A(4), 157-160.
Moon, F. C., Holmes, W. T. (1985), "The fractal dimension of the two-well potential strange attractor", Physica 17D, 99-108.
Moon, F. C., Li, G.-X. (1985), "Fractal basin boundaries and homoclinic orbits for periodic motion in a two-well potential", Phys. Rev. Lett., 55(14), 1439-1442.
Moon, F. C., and Holmes, P. J. (1979), "A magnetoelastic strange attractor," J. Sound Vib., Vol. 65, No. 2, pp. 275-296; Moon, F. C., and Holmes, P. J., A Magnetoelastic strange attractor," J. Sound Vib., 69, No. 2, p. 339.
Moon, F. C. (1980), "Experiments on chaotic motions of a forced nonlinear oscillator: Strange attractors," ASME Journal of Applied Mechanics, 47, pp. 638-44.
Moon, F. C. (1984), Fractal boundary for chaos in a two-state mechanical oscillator, Phys. Rev. Lett., 53, No. 60, pp. 962-964.
Newhouse, S., Ruelle, D., and Takens, F. (1978), "Occurrence of strange axiom A attractors near quasi periodic flows on T^m, $m \geq 3$, Commun. Math. Phys., 64, 35-40.
Nayfeh, A. H., and Mook, D. T. (1979), Nonlinear Oscillations, J. Wiley and Sons.
Packard, N. H., Crutchfield, J. P., Farmer, J. D., and Shaw, R. S. (1980), "Geometry from a time series", Phys. Rev. Lett., 45, 712.
Pomeau, Y., Mannville, P. (1980), "Intermittent transition to turbulence in dissipative dynamical systems", Commun. Math. Phys., 74, 189-197.
Rollins, R. W., and Hunt, E. R. (1982), "Exactly solvable model of a physical system exhibiting universal chaotic behavior", Phys. Rev. Lett. 49 (18), 1295-1298.
Schuster, H. G. (1984), Deterministic Chaos, Physic-Verlag GmbH, Weinheim (FRG).
Shaw, S., Holmes, P. J. (1983), "A periodically forced piecewise linear oscillator", J. Sound and Vibration, 90(1), 129-155.
Swinney, H. L. (1983), "Observations of order and chaos in nonlinear systems," in Order and Chaos, Campbell and Rose, eds., North-Holland, Amsterdam, 3-15.
Ueda, Y. (1979), "Randomly transitional phenomena in the system governed by Duffing's equation," J. Statistical Physics, 20, 181-196.
Wolf, A. (1984), "Quantifying chaos with Lyapunov exponents", Nonlinear Science: Theory and Application.
Wolf, A., Swift, J. B., Swinney, H. L., Vasano, J. A. (1985), "Determining Lyapunov exponents from a time series", Physica 16D, 285-317.

CHAOTIC AND REGULAR MOTION IN NONLINEAR VIBRATING SYSTEMS

W. Szemplinska-Stupnicka
Institute of Fundamental Technological Research, Warsaw, Poland

1. Introduction

Studies of phenomena arising in nonlinear oscillators are often modelled by an equation of the form

$$\ddot{x} + g(\dot{x}) + \psi(x) = f \cos \omega t$$

where $g(\dot{x})$ and $\psi(x)$ are approximated by finite Taylor series, and $g(\dot{x})$ represents a dissipative term. Such a system has an extensive literature. A,now classical approach to the study of the system behaviour, such as that presented in the popular book by Hayashi [8], is the theoretical analysis based on approximate analytical methods with experimental verification employing electric circuits or electronic computers. In these studies the system is assumed to tend to steady-state oscillation when started with any initial conditions and steady-state solutions are often the main point of interest. Approximate analytical solutions describing various types of resonances and analysis of local stability of the solutions and their domains of attraction provided us with a great deal of knowledge about the system behaviour. The results show a variety of nonlinear phenomena such as: principal, sub, ultra and subultra harmonic resonances and jump phenomena associated with stability limits on resonance curves, which seem to leave no room for any irregular, random-like and unpredictable solutions in the deterministic systems.

Although chaotic motion in simple deterministic dynamic systems have attracted a great deal of attention in the last decade, results showing "strange attractors" in as classical a vibrating system as that governed by Duffing equation,were a great surprise [26-28].

It is pretty obvious that direct applications of the approximate theory of nonlinear vibrations to theoretical study of chaotic motion is impossible and so the return to qualitative topological methods seemed to be the only alternative. Nevertheless one might be tempted to seek a link between the phenomena of nonlinear resonances determined by low order approximate solutions and irregular solutions obtained by computer simulation in order to see the chaotic zones against the background of the classical concepts of resonance curves, stability limits, and jump phenomena. Results on chaotic behaviour obtained by computer simulation allow us to make observations,which make such an idea an appealing one: one can readily notice that:

- chaotic motion appears as a transition zone between sub or subultra mT-periodic resonance and the principal T-periodic resonance,
- chaotic motion often borders and coexists with periodic motion, the motion which can be described by low order approximate solutions.

The idea of interpreting and studying regions of chaotic motion from the point of view of the approximate theory of nonlinear vibrations brings a great number of interesting questions:

- where are the chaotic motion zones located relative to the known phenomena of principal and subharmonic resonances?
- how does chaotic motion develop from periodic solutions if the periodic motion is described by a low order approximate solution and the approximate theory of nonlinear vibrations does not leave room for irregular solutions?
- whether and how routes to chaotic motion can be approximately described with the use of concepts and mathematical language of the approximate theory of nonlinear vibrations?
- can the low order approximate solutions make foundations for approximate criteria of system parameter critical values – the parameters for which one might expect chaotic behaviour?

It is an attempt of this work to seek answers to some of the questions. A consistent use of mathematical tools and concepts commonly accepted in the approximate theory of nonlinear vibrations,and their adaptation to an analysis of routes to chaotic motion,is one of the main points of interest.

The following three oscillators are studied in detail:

1. An oscillator with a single equilibrium position and unsymmetric elastic characteristic governed by an equation of the form:

$$\ddot{x} + \gamma\dot{x} + \mathcal{H}_1 x^2 + \mathcal{H}_2 x^3 = f \cos \omega t ;$$

2. An oscillator with a single equilibrium position and symmetric elastic nonlinearity governed by Duffing's equation :

$$\ddot{x} + \gamma x + \dot{x}^3 = f \cos \omega t \; ;$$

3. An oscillator with three positions of equilibrium governed by an equation of the form:

$$\ddot{x} + \gamma \dot{x} - \frac{1}{2}(1 - x^2)x = f \cos \omega t \; ;$$

In the dissipative systems to be considered it is assumed that any solution started with initial conditions $x(0)$, $\dot{x}(0)$ tends, after some transient, to a steady-state oscillation: regular oscillations – periodic or to chaotic motion.

To observe the steady-state oscillations and then identify them, four descriptions will be employed:

- time history $x \equiv x(t)$
- phase portrait $x \equiv x(\dot{x})$
- frequency spectrum $x(nT) \equiv x[\dot{x}(nT)]$;
- Poincare map.

The first three are standard and require here no further elaboration. Poincare map, which plays an important role in investigations of chaotic motion,is a plot in the phase plane at chosen discrete instants of time. For the nonautonomous systems considered here,the discrete instants of time are separated by equal time intervals corresponding to a period of excitation. The sampling time is chosen to be equal to the period of excitation term:

$$T = 2\pi/\omega$$

so that periodic motion of the period mT is mapped as m points on the $x(nT) - \dot{x}(nT)$ plane, $n = 1,2,3,\ldots$.

Regular periodic and almost periodic oscillations

In Figs. 1.1.a and b two examples of periodic motion are illustrated. Phase portraits show closed curves and two or three points, corresponding to 2T and 3T-periodic motion respectively, are seen in Poincare maps. A complex form of the phase portrait in Fig. 1.b is due to a high number of harmonic components involved in the solution $x(t)$. While only 3 harmonic components are visible in the case (a), the 3T-periodic motion is characterized by five components. The two types of periodic motion can be described by simple, closed form formulae:

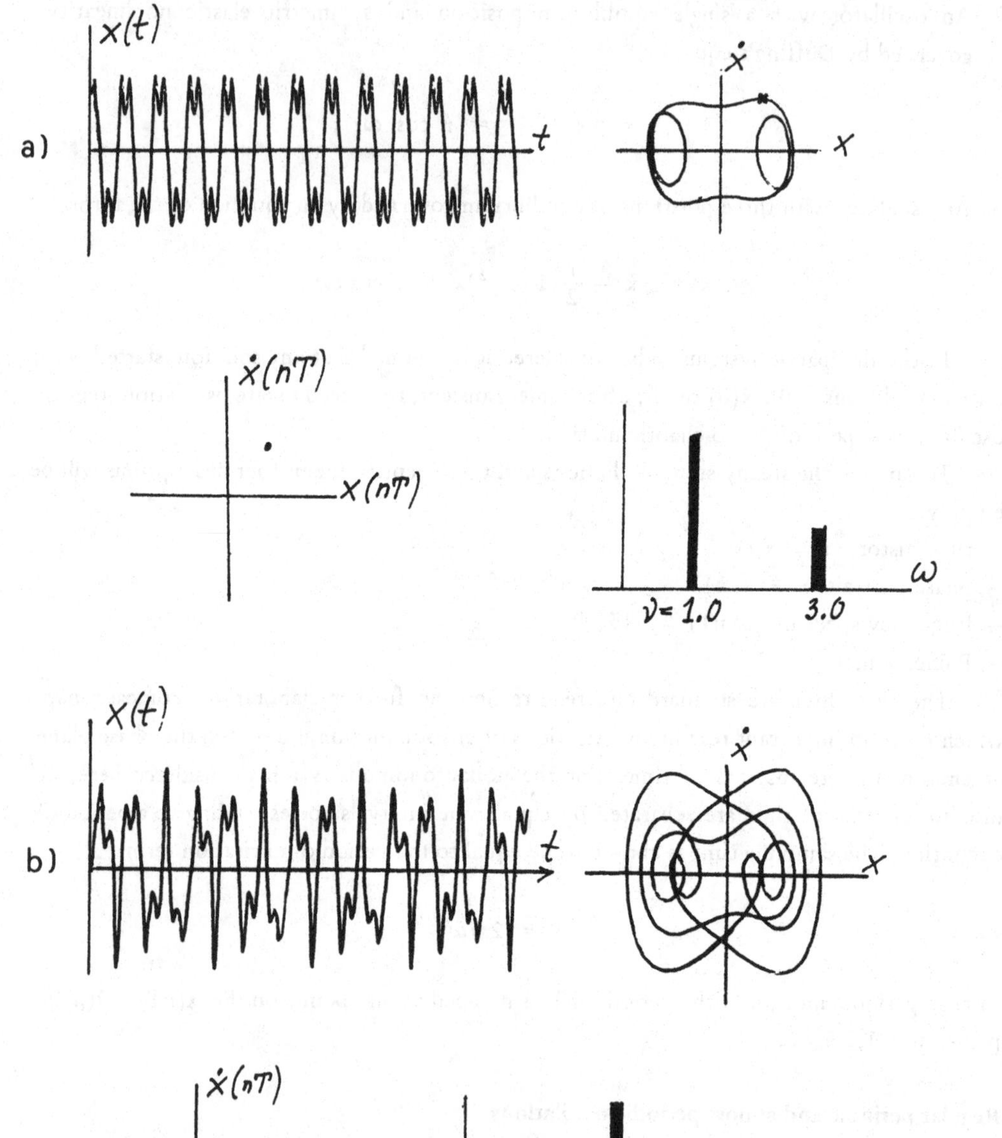

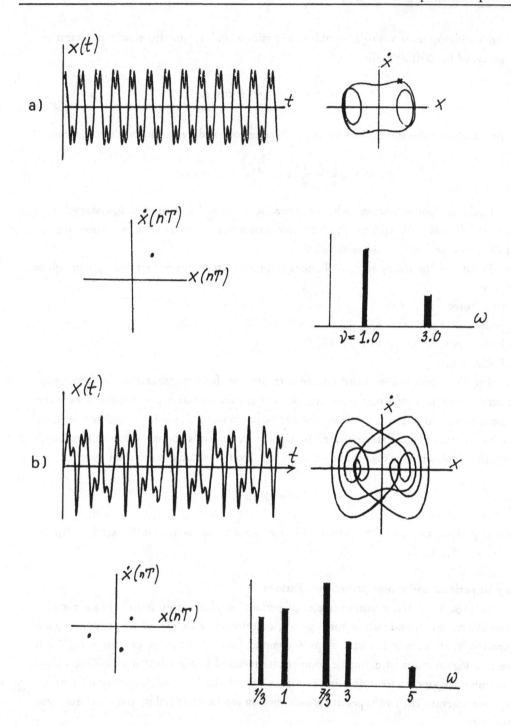

Fig. 1.1. Four descriptors of periodic motion: (a) T-periodic, (b) 3T-periodic motion.

$$x(t) = A_0 + A_1 \cos(\omega t + \vartheta_1) + A_{1/2} \cos(\frac{1}{2}\omega t + \phi) = x(t + 2T) ;$$

and

$$x(t) = \sum_{p=1,3,5} Ap \cos(p\omega t + \vartheta_p) + \sum_{p=1,7} A_{p/3} \cos(p \frac{\omega}{3} t + \vartheta_{p/3}) \equiv x(t + 3T) ;$$

respectively.

Figs. 1.2 and 1.3 show four characteristics of *almost-periodic oscillations.* Fig. 2 illustrates the simplest case of the almost-periodic motion when only two harmonic components with irrational frequencies are involved, so that the signal can be written as

$$x(t) = A_1 \cos \omega t + A_2 \cos \nu t ,$$

where

$$\frac{\nu}{\omega} \neq \frac{n_1}{n_2} , \qquad n_1, n_2 = 1,2,3, \ldots$$

and $T = \dfrac{2\pi}{\omega}$ is the sampling time of the Poincare map.

For almost-periodic motion, the phase portrait does not show a closed curve. After a sufficiently long time interval, trajectories fill up a section of phase plane, while a Poincare map shows an elliptically shaped closed curve. In more general case with two irrational frequencies written as

$$x(t) \ \sum_{n=1,2,3\ldots} A_{1n} \cos(n\,\omega t + \phi_n) + \sum_{n=1,2,3\ldots} A_{2n} \cos(n\nu t + \vartheta_n) ,$$

$$\frac{\nu}{\omega} \neq \frac{n_1}{n_2} , \qquad n_1 , n_2 = 1,2,3, \ldots$$

the Poincare map will still form a smooth closed curve.

When more than two irrational frequencies are involved in x(t), a plot on Poincare map becomes more complex: an infinite set of points spread uniformly in a section of phase plane will appear. This is shown in Fig. 1.3 where motion consisting of three harmonic components are illustrated:

$$x(t) = A_1 \cos \omega_1 t + A_2 \cos \omega_2 t + A_3 \cos \omega t ,$$

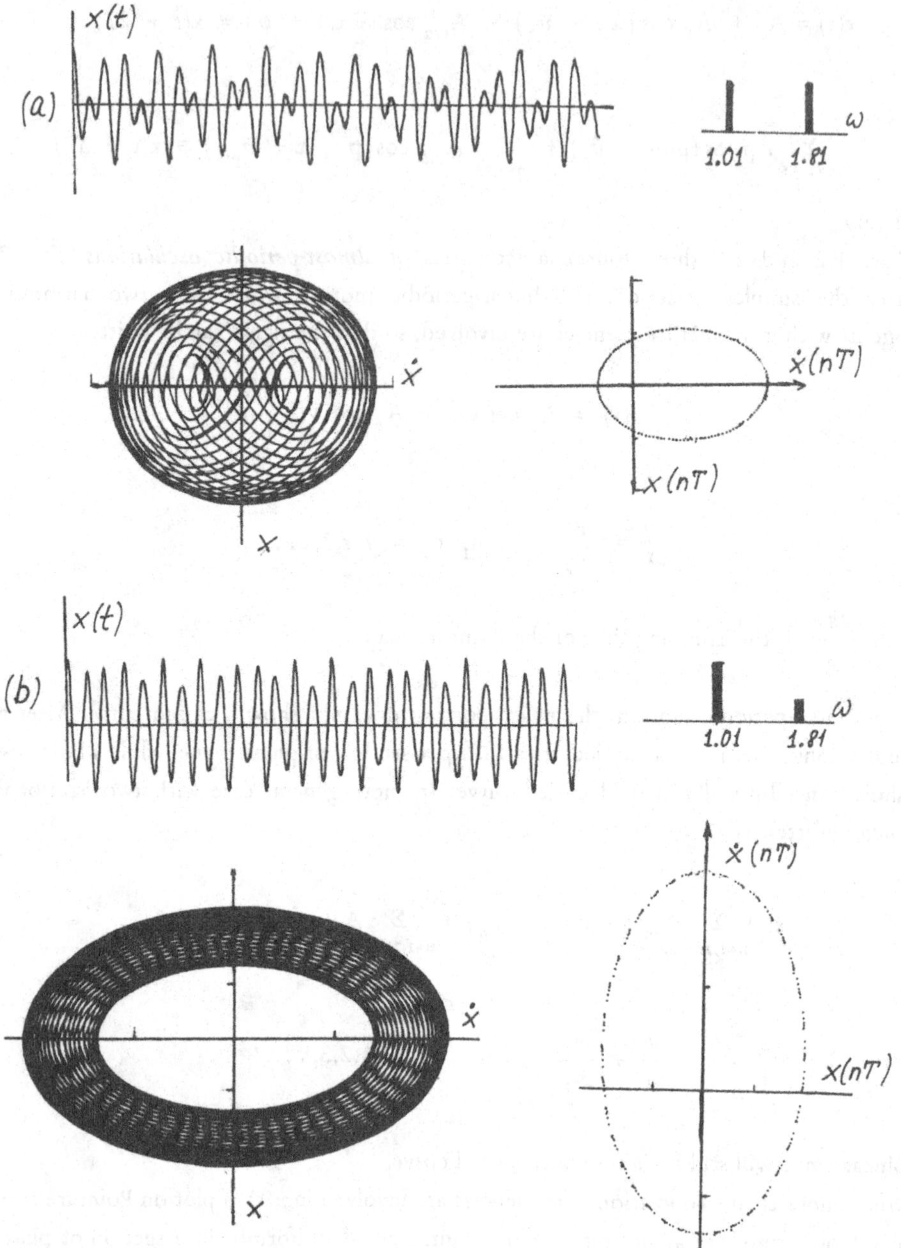

Fig. 1.2. Four descriptors of almost-periodic motion – two irrational frequencies.

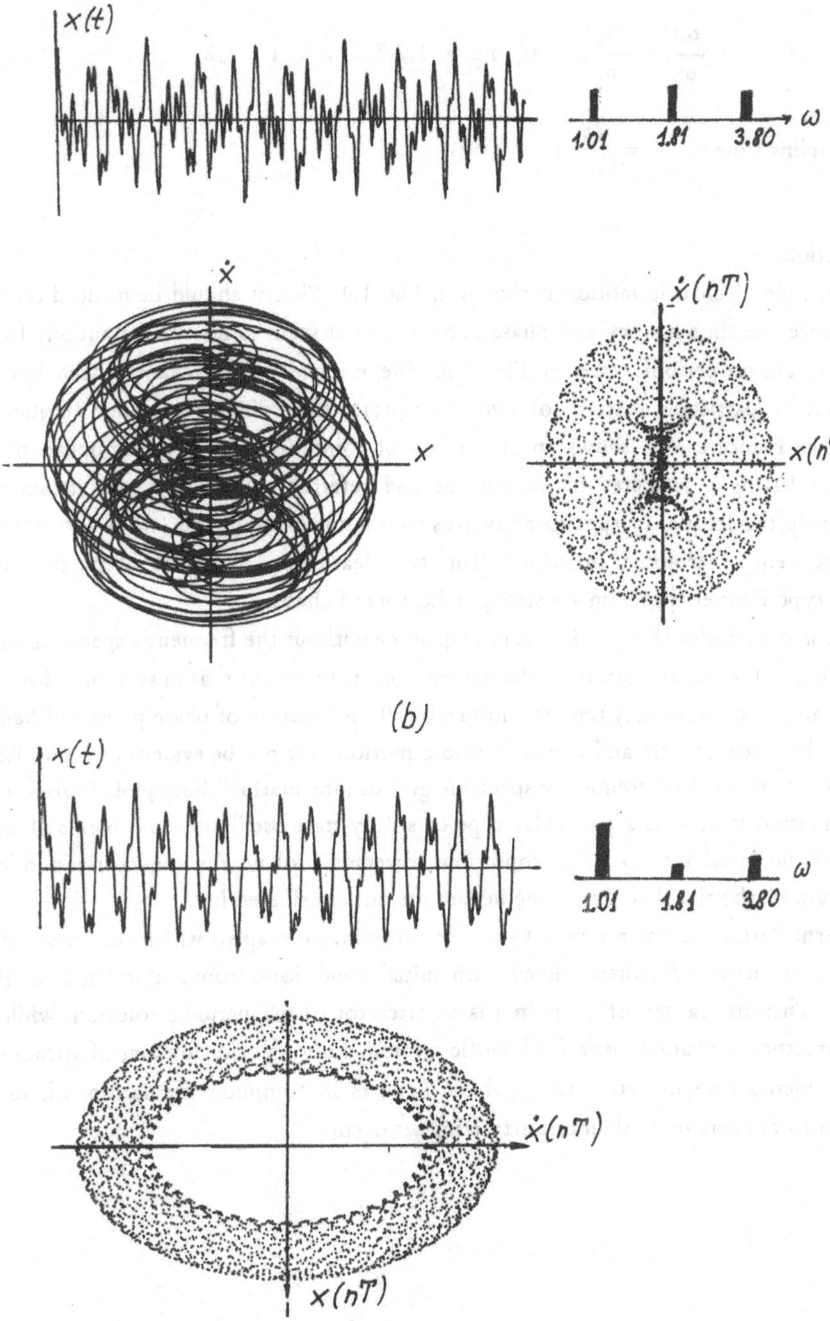

Fig. 1.3. Four descriptors of almost-periodic motion – three irrational frequencies.

where

$$\frac{\omega i}{\omega} \neq \frac{n_1}{n_2}, \quad n_1, n_2 = 1,2,3,\ldots, \quad i = 1,2;$$

and the sampling time is $T = \frac{2\pi}{\omega}$;

Chaotic motion

An example of chaotic motion is shown in Fig. 1.4. First it should be noticed that at the first glance the time history and phase portrait do not seem to differ substantially from those of the almost-periodic case in Fig. 1.3. The essence of the chaotic behaviour is characterized by particular features of two descriptors: the Poincare map and frequency spectrum. The Poincare map shows an infinite set of points arranged in what appears to be parallel lines having a property of Cantor set and referred to as a "strange attractor". Simultaneously, the frequency spectrum involves continuous components (while the input is a single frequency harmonic function). The two features – "strange attractor" and continuous type Fourier spectrum are strong indicators of chaos.

It is worth to notice that the Poincare map alone without the frequency spectrum may not be sufficient for identification of chaotic motion. It turns out that in systems with low damping strange attractors may tend to uniformly fill up a section of phase plane and hence a difference between chaotic and almost-periodic motion may not be evident (see Fig. 1.5).

The Poincare map and frequency spectrum give us information about global properties of chaotic motion treated as a particular type of steady-state oscillations. Problems of high sensitivity of the signal x(t) to initial conditions, divergence of nearby trajectories and loss of information on the signal at given time instant are not considered here.

The term "attractor" is referred to a plot on Poincare map, to which our dissipative system tends (is attracted), when started with initial conditions from a domain D of the phase-plane. Therefore, a set of m points is an attractor of mT-periodic solution, while a "strange attractor" is characteristic for chaotic motion. Although the domains of attraction are not considered here in detail, the problem appears in computer simulation, where at given system parameters more than one attracting set occurs.

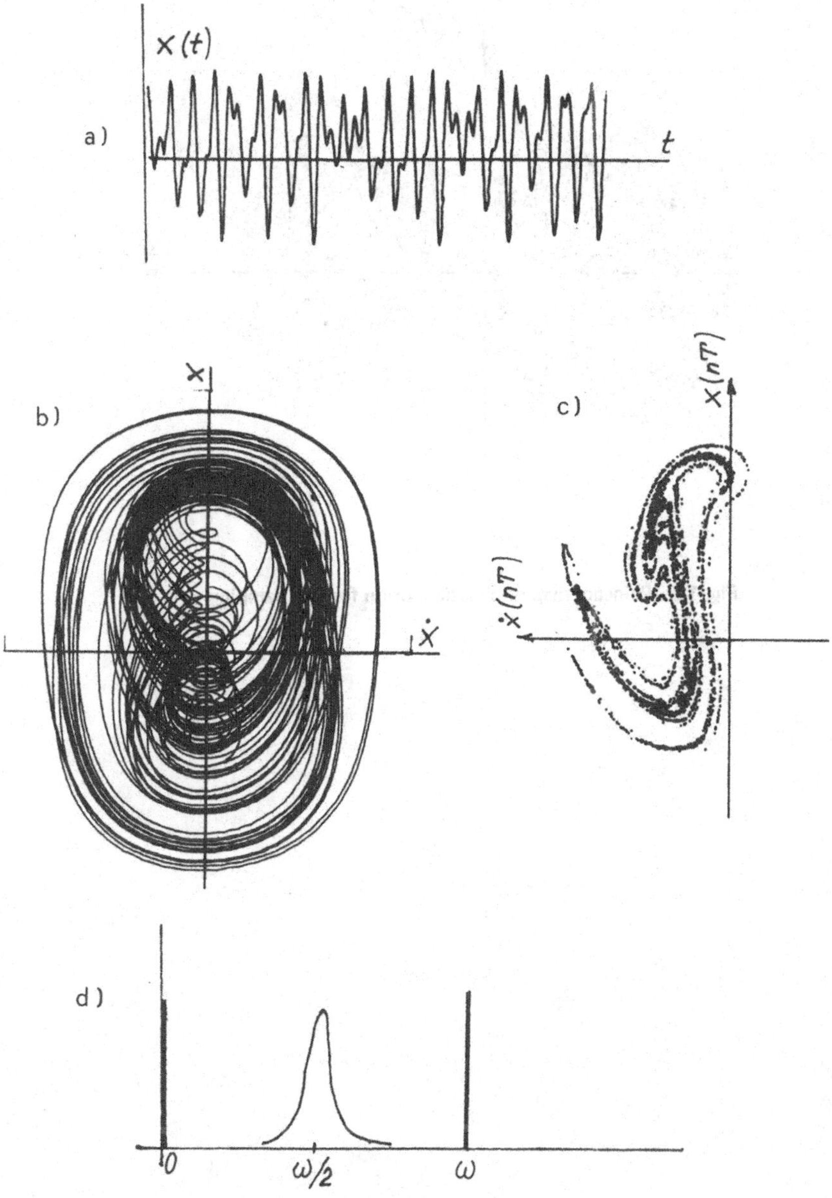

Fig. 1.4. Four descriptors of an example of chaotic motion.

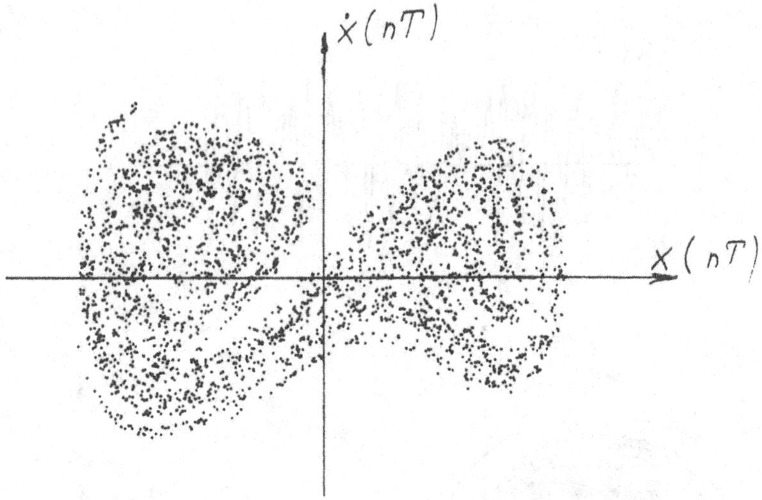

Fig. 1.5. Poincare map of chaotic motion for low damping.

2. An Oscillator with Unsymmetric Elastic Nonlinearity – The 2T-Subharmonic Resonance and its Transition to Chaos

In this section we consider an oscillator with a single equilibrium position governed by an equation of the form:

$$\ddot{y} + \gamma\dot{y} + \omega_0^2\, y + \psi(y) = f \cos \omega t$$

$$\psi(y) \neq -\psi(-y) \tag{2.1.a}$$

$$\psi(0) = 0\,; \quad \psi(y) \neq 0 \ \text{ at } \ y \neq 0\,;$$

where $\psi(y)$ is approximated by two term Taylor expansion

$$\psi(y) = \mathcal{K}_1\, y^2 + \mathcal{K}_2 y^3\ ; \tag{2.1.b}$$

The behaviour of this system has been studied extensively by approximate analytical methods including the phenomena of sub, ultra and subultraharmonic resonances, jump phenomena, problems of stability of various approximate solutions and their domains of attraction. These investigations predicted solutions with satisfactory accuracy using low order approximate solutions ,which were confirmed by experiments ,or by computer simulation (see e.g. [8,23,25]). Systems with piece-wise nonsymmetric characteristics treated by various analytical and numerical techniques showed similar properties (e.g. [1,12]).

It is known that the 2T and 3T subharmonic resonances are those,which appear at relatively low values of the forcing parameter f and coefficients $\mathcal{K}_1$, $\mathcal{K}_2$. They can be described by a low order approximate solution as

$$y(t) = y(t + 2T) = B_0 + B_{1/2} \cos\left(\frac{\omega}{2} t + \phi_2\right) + B_1 \cos(\omega t + \vartheta)\,; \tag{2.2.a}$$

$$T = 2\pi/\omega\ ;$$

and

$$y = y(t + 3T) = B_1 \cos(\omega t + \vartheta_1) + B_{1/3} \cos\left(\frac{\omega}{3} t + \phi_3\right); \tag{2.2.b}$$

and occur in the neighbourhood of frequency $\nu \approx 2\omega_0$ and $\nu \approx 3\omega_0$ respectively. The subharmonic resonances coexist with the principal one described by the first approximate formulae

$$y(t) = y(t + T) = C_0 + C_1 \cos(\omega t + \vartheta) , \qquad (2.2.c)$$

and the response of the system jumps from one type of steady-state oscillations to another at points of vertical tangent on resonance curves , i.e.,the classical stability limits. Thus the approximate theory of nonlinear vibrations says, that the steady-state vibrations can be separated in time by transient states only. The transients can take place when a stability limit of a particular type of resonance is reached, or when the system is started with initial conditions different from those coinciding with one of possible steady-state vibrations.

However in 1979, a distinctly new type of the steady-state motion – called "randomly transitional phenomena" or "chaotic motion" – was reported by Ueda [26, 27]. It has become clear that our knowledge about the properties and behaviour of the system is far from being complete. While the chaotic behaviour in systems having three positions of equilibrium studied extensively by theoretical and experimental methods [2–7, 13] can be intuitively explained by some physical arguments, the arguments fail in the system where $y = 0$ is the only one rest point [19].

The hitherto results on the chaotic behaviour of the system allow us to make an observation that the phenomena is associated with a transition from the 2T-subharmonic resonance to the principal one, that is, to the response of the period of excitation. The observation gives an appealing idea to try to find a link between the regular periodic solution (2.2a-c), and the chaotic motion described in terms of "strange attractors", to see the chaotic zone against the background of the classical resonance curves in connection with the concepts of the stability limits and jump phenomena [21].

2.1. Harmonic solution, its local stability and period doubling bifurcation

We consider the system governed by eqs. (2.1a.b) where, for the sake of simplicity, it is assumed,

$$\omega_0^2 = 3\sqrt[3]{P_0^2}$$

$$\mathcal{K}_1 = 3\sqrt[3]{P_0} \ , \quad \mathcal{K}_2 = 1 ;$$

so that eq. (2.1) is reduced into the form:

$$\ddot{x} + \gamma\dot{x} + x^3 = P_0 + f \cos \nu t \ ; \tag{2.3}$$

where

$$x = y + \sqrt[3]{P_0}$$

Eq. (2.3) was examined by Ueda and the strange attractors were found at $\omega = 1.0$; $\gamma = 0.05$; $f = 0.16$; $P_0 = 0.030$ and 0.045 [27].

In the first step of the theoretical analysis we consider the solution with period of the exciting force, called later the "T-periodic solution", in the form:

$$x_0^{(1)}(t) = C_0 + C_1 \cos(\omega t + \vartheta) \equiv x(t + T) \tag{2.4a}$$

Since we deal with strongly nonlinear systems,' any simplifications inherent to the perturbation methods should be avoided and hence we will consistently make use of the harmonic balance method [8].

On substitution eqs. (2.4a) into eqs. (2.3) and equating coefficients of $\cos(\omega t + \vartheta)$, $\sin(\omega t + \vartheta)$, and constant term separately to zero, a set of algebraic equations for the unknown C_0, C_1, ϑ is obtained

$$\begin{aligned} -C_1\,\omega^2 + 3/4 C_1^3 + 3\,C_0^2\,C_1 &= f \cos \vartheta \ , \\ -\gamma \nu C_1 &= f \sin \vartheta \ , \\ C_0^3 + 3/2\, C_0\, C_1^2 &= P_0 \ ; \end{aligned} \tag{2.4b}$$

On solving eqs. (2.4b) by a numerical procedure, resonance curves $C_0 \equiv C_1(\omega)$ $C_1 \equiv C_1(\omega)$ are obtained, (Fig. 2.1).

To examine local stability of the solution (2.4a) thus determined, we consider a disturbed solution,

$$x(t) = x_0^{(1)}(t) + \delta x \ , \tag{2.5a}$$

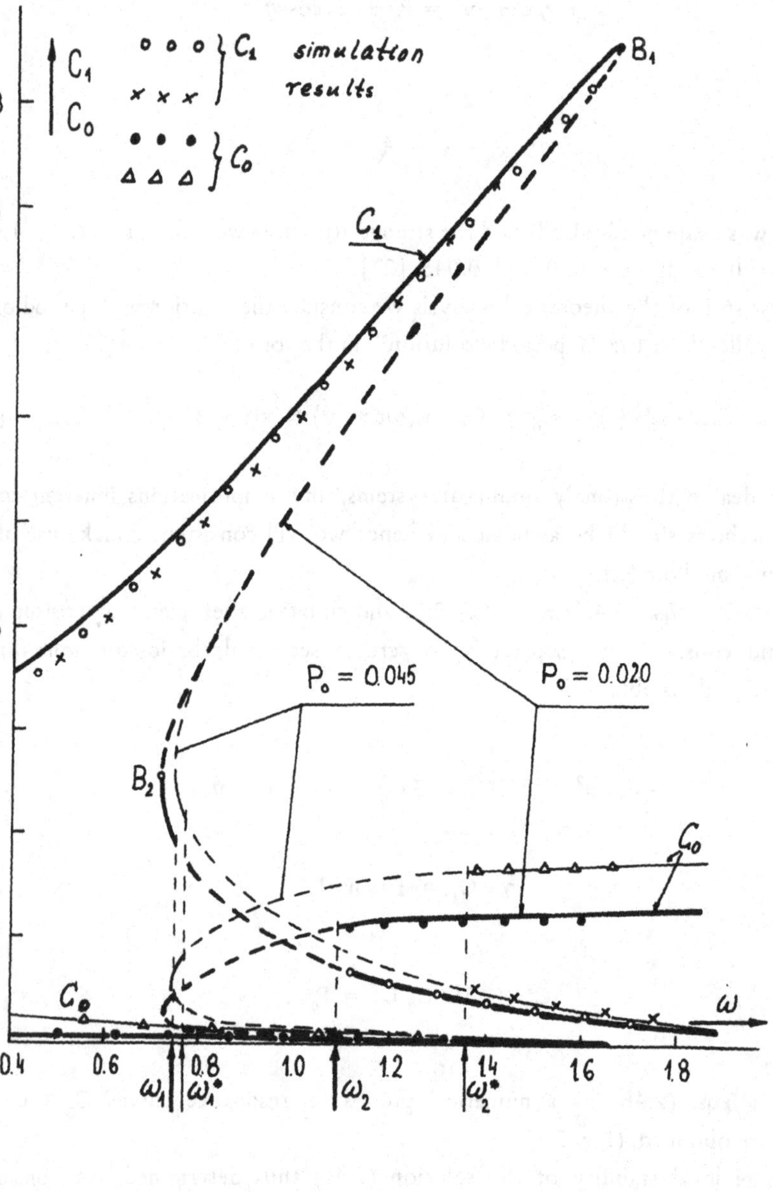

Fig. 2.1. Resonance curves of harmonic plus constant term solution; $f = 0.16$, $\gamma = 0.05$.

and on inserting it into eqs. (2.3) and taking into account eqs. (2.4b) we arrive at the variational equation:

$$\delta\ddot{x} + \gamma\delta\dot{x} + [3x_0^{(1)^2}(t)]\,\delta x + 3x_0^{(1)}(t)\delta x^2 + \delta x^3 = 0\;; \tag{2.5b}$$

The question of local stability is answered on ignoring higher power terms and considering the linear variational equation,

$$\delta\ddot{x} + \gamma\delta\dot{x} + [3x_0^{(1)^2}(t)]\delta x = 0\;; \tag{2.5c}$$

Then inserting eqs. (2.4a) and expanding $x_0^2(t)$ into the Fourier series yields:

$$\delta\ddot{x} + \gamma\delta\dot{x} + \delta x[\lambda_0 + \lambda_1\cos(\omega t + \vartheta) + \lambda_2\cos(2\omega t + 2\vartheta)] = 0\;;$$

$$\lambda_0 = 3C_0^3 + 3/2\,C_1^2\,,$$

$$\lambda_1 = 6\,C_0\,C_1\,, \tag{2.5d}$$

$$\lambda_2 = \frac{3}{2}\,C_1^2$$

Thus we arrive at the Hill's equation studied extensively in [8] by the aid of the harmonic balance method. To examine the behaviour of the disturbance $\delta x(t)$ and hence the stability of the harmonic solution we will make use of the method and results presented in the book. We readily see that the classical first order unstable region known in Duffing's equation, which is due to the term $\lambda_2\cos[2(\omega t + \vartheta)]$ takes place at

$$\omega \approx \sqrt{\lambda_0}\,,$$

and the stability limits coincide with points of vertical tangents on the resonance curves. In Fig. 2.1 the region is between points B_1, B_2 and is denoted by a dashed line.

However, due to the constant term C_0, a parametric term of the frequency ω –, $\lambda_1\cos(\omega t + \vartheta)$ appears in the variational equation. Consequently, the lowest order unstable region is that which occurs close to

$$\omega \approx 2\sqrt{\lambda_0}\ , \tag{2.6a}$$

and this type of instability needs a particular attention.

The first approximate solution in the unstable region of the type is,

$$\delta x(t) = e^{\varepsilon t}\, b_{1/2} \cos\left(\frac{\omega}{2} t + \phi\right), \tag{2.6b}$$

where ϵ is real and positive. At the stability limit $\epsilon = 0$ and hence

$$\delta x(t) = b_{1/2} \cos\left(\frac{\omega}{2} t + \phi\right); \tag{2.6c}$$

The form of solution (2.6b) suggests that there is a possibility of a build-up of the harmonic component of the frequency $\omega/2$, that is – bifurcation from the T-periodic solution (2.4a) to 2T-periodic solution. To determine boundaries of the unstable region on the resonance curves $C_0(\omega)$, $C_1(\omega)$ we insert the assumed solution (2.6c) into the variational equation (2.5d) and apply the harmonic balance method. Conditions of nonzero solutions for $b_{1/2}$ lead us to the following criterion to be satisfied at the stability limits:

$$\underset{\varepsilon=0}{\Delta(\omega^2)} = \left(\lambda_0 - \frac{\omega^2}{4}\right)^2 + \gamma^2 \frac{\omega^2}{4} - \frac{1}{4}\lambda_1^2 = 0\,; \tag{2.6d}$$

Inside the unstable region, that is at $\epsilon > 0$ the determinant is negative,

$$\underset{\varepsilon>0}{\Delta(\omega^2)} < 0 \tag{2.6e}$$

On utilizing eqs. (2.6d.e) the unstable region was evaluated and denoted in Fig. 2.1. It lies on the lower, nonresonant branch of $C_1(\omega)$ and associated upper branch of $C_0(\omega)$, between frequencies ω_1 and ω_2.

To answer the question of period doubling bifurcation [9] at the critical point $\omega = \omega_1$ with increasing frequency and at $\omega = \omega_2$ with decreasing frequency, we turn back to the complete nonlinear variational equation (2.5b) and examine the existence and stability of the steady-state solution (2.6c) in a small neighbourhood of the two critical points. On inserting eqs. (2.4a) the complete variational equation takes the form:

$$\delta\ddot{x} + \gamma\delta\dot{x} + \delta x[\lambda_0 + \lambda_1 \cos\Theta + \lambda_2 \cos 2\Theta] +$$

$$+ \delta x^2(3C_0 + 3C_1 \cos\Theta) + \delta x^3 = 0 ,$$

$$\Theta = \omega t + \vartheta ,$$

$$\lambda_0 = 3C_0^2 + 3/2\, C_1^2 , \tag{2.7a}$$

$$\lambda_1 = 6\, C_0\, C_1 , \quad \lambda_2 = 3/2\, C_1^2 ;$$

Values of C_0 and C_1 are assumed to be equal to those at $\omega = \omega_1$ or $\omega = \omega_2$ respectively. Since only a small neighbourhood of the critical points is to be considered and we are not only interested in the steady state solution (2.6c) but also in its local stability, we make use of the perturbation method. Therefore we introduce a small parameter μ into eqs. 2.7a and rewrite it into:

$$\delta\ddot{x} + \frac{\nu^2}{4}\,\delta x + \mu(3\check{C}_0 + 3\check{C}_1 \cos\Theta)\delta x^2 + \mu^2[\bar{h}\delta x +$$

$$\delta x(\Delta\omega + \bar{\lambda}_1 \cos\Theta + \bar{\lambda}_2 \cos 2\Theta)] + \delta x^3 = 0 , \tag{2.7b}$$

where

$$\mu^2\Delta\omega = \lambda_0 - \frac{\nu^2}{4} , \quad \mu^2\check{h} = h, \quad \mu^2\check{\lambda}_{1,2} = \lambda_{1,2} ,$$

Then the 2T-periodic solution is sought in the form of power series in μ :

$$\delta x(t) = b_1 \cos\left(\frac{\nu}{2}t + \phi\right) + \mu u_1(t) + \mu^2 u_2(t) ,$$

$$\frac{db_1}{dt} = \mu E_1 + \mu^2 D_1 , \tag{2.7c}$$

$$\frac{d\phi}{dt} = \mu E_2 + \mu^2 D_2 ;$$

On applying the classical perturbation technique we easily arrive to the following result:

$$E_1 = E_2 = 0$$

$$\delta x(t) = b_1 \cos(\frac{\nu}{2} t + \phi) + b_0 + b_2 \cos(\nu t + 2\phi);$$

$$b_0 = -\frac{6C_0 b_1^2}{4\lambda_0}; \quad b_2 = \frac{C_0 b_1^2}{2\lambda_0};$$

$$\frac{db_{1/2}}{dt} = \frac{-b_{1/2}}{\omega}[\frac{\lambda_1}{2} \sin(\vartheta - 2\phi) + \gamma \frac{\omega}{2}] \equiv D_1(b_{1/2}, \phi);$$

$$\frac{d\phi}{dt} = \frac{1}{\omega}[\lambda_0 - \frac{\omega^2}{4} + \frac{\lambda_1}{2} \cos(\vartheta - 2\phi) + \beta\, b_{1/2}^2] \equiv D_2(b_{1/2}, \phi); \quad (2.7d)$$

The steady-state solution is determined by conditions,

$$\frac{db_{1/2}}{dt} = \frac{d\phi}{dt} = 0,$$

which yield equations for $b_{1/2}$ ϕ:

$$\lambda_0 - \frac{\omega^2}{4} + \frac{\lambda_1}{2} \cos(\vartheta - 2\phi) + \beta\, b_{1/2}^2 = 0,$$

$$\frac{\lambda_1}{2} \sin(\vartheta - 2\phi) + \gamma \frac{\omega}{2} = 0 \quad (2.8a)$$

$$\beta = \frac{3}{4}(1 - \frac{10\, C_0^2}{\lambda_0})$$

Then on eliminating the angle $\vartheta - 2\phi$ we obtain:

$$\Delta(\omega^2, b_{1/2}) = (\lambda_0 - \frac{\omega^2}{4} + \beta\, b_{1/2}^2)^2 + \gamma^2 \frac{\omega^2}{4} - \frac{\lambda_1^2}{4} = 0\,; \tag{2.8b}$$

To find bifurcation diagrams in terms of the parameter

$$\mu_1 = \omega^2 - \omega_1^2 \quad \text{in the vicinity of } \omega_1\,, \text{ and}$$

$$\mu_2 = \omega_2^2 - \omega^2 \quad \text{in the vicinity of } \omega_2\,,$$

we expand the determinant (2.8b) into a power series with the result:

$$\Delta(\omega^2, b_{1/2}) = \Delta(\omega^2, 0) + \frac{\partial\Delta}{\partial\omega^2}(\omega^2 - \omega_{1,2}^2) + \frac{\partial\Delta}{\partial b_{1/2}^2}\, b_{1/2}^2 = 0\,; \tag{2.8c}$$

By virtue of eqs. (2.6d)

$$\Delta(\omega^2, 0) = 0\,,$$

and eventually the amplitude of the bifurcating solution is found to be,

$$\frac{4}{3}\,\beta_{(1)}\, b_{1/2}^2 = \frac{1}{3}\left[1 - \frac{\gamma^2}{2(\lambda_0 - \frac{\omega_1^2}{4})}\right]\mu_1\,,$$

or (2.8d)

$$-\frac{4}{3}\,\beta_{(2)}\, b_{1/2}^2 = \frac{1}{3}\left[1 + \frac{\gamma^2}{2(\frac{\omega_2^2}{4} - \lambda_0)}\right]\mu_2\,,$$

where

$$\beta_{(1)} > 0 \qquad \text{and} \qquad \beta_{(2)} > 0\,;$$

Thus we arrive at the typical bifurcation diagrams sketched in Fig. 2.2.

To examine local stability of the steady-state solution,

$$\delta x = b_{1/2}\cos\left(\frac{\omega}{2}t + \phi\right) + \mu\, u_2(t, b_{1/2}, \phi), \tag{2.9a}$$

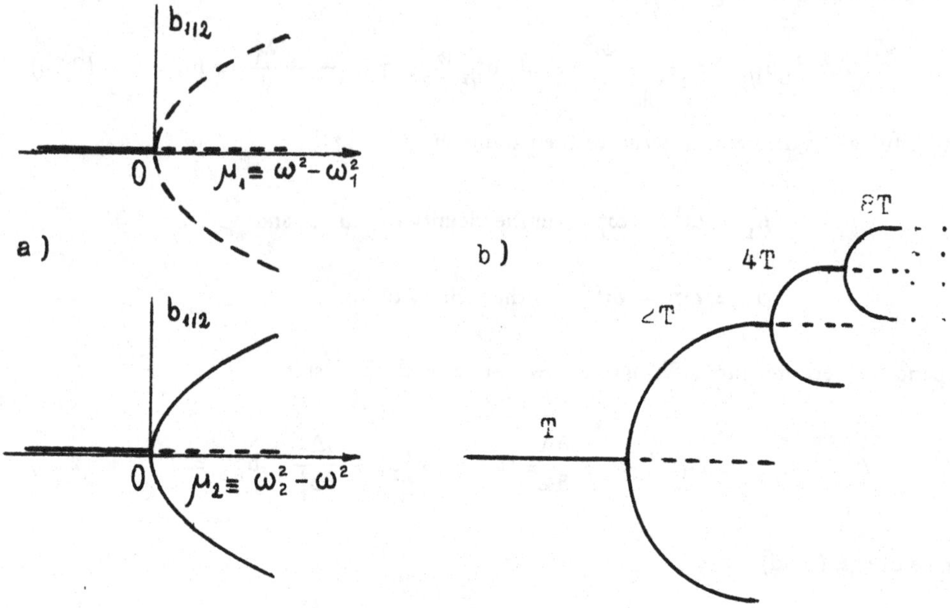

Fig. 2.2. Period doubling bifurcation diagram.

we turn back to eqs. (2.7d) and add small disturbances to $b_{1/2}$, ϕ determined by eqs. (2.8a) to obtain linear variational equations:

$$\frac{d\delta b_{1/2}}{dt} = \frac{\partial D_1}{\partial b_{1/2}} \delta b_{1/2} + \frac{\partial D_1}{\partial \phi} \delta\phi ,$$

$$\frac{d\delta\phi}{dt} = \frac{\partial D_2}{\partial b_{1/2}} \delta b_{1/2} + \frac{\partial D_2}{\partial \phi} \delta\phi ; \qquad (2.9b)$$

Conditions of nontrivial particular solution,

$$\delta b_{1/2}(t) = B_1 e^{\epsilon t} ,$$

$$\delta\phi(t) = B_2 e^{\epsilon t} , \qquad (2.9c)$$

lead to a characteristic determinant of the form,

$$\begin{vmatrix} -\varepsilon & \dfrac{+b_{1/2}\lambda_1}{\omega}\cos(\vartheta-2\phi) \\ 2\beta\omega\, b_{1/2} & \dfrac{\lambda_1}{\omega}\sin(\vartheta-2\phi) \end{vmatrix} = 0, \tag{2.9d}$$

which, by virtue of eqs. (2.8a) is reduced to,

$$\varepsilon^2 + \gamma\varepsilon - \frac{16\,\beta_{1,2}}{9}\frac{b^2_{1/2}}{\omega^2}\left(\frac{\omega^2_{1,2}}{4} - \lambda_0\right) = 0; \qquad \gamma > 0. \tag{2.9e}$$

On making use of the Routh-Hurwitz criterion [8] we see that the solution (2.9c) is stable if

$$-\frac{16\,\beta_{1,2}}{9}\frac{b^2_{1/2}}{\omega^2}\left(\frac{\omega^2_{1,2}}{4} - \lambda_0\right) > 0;$$

From eqs. (2.6d) we learn that,

$$\beta_1\left(\frac{\omega_1^2}{4} - \lambda_0\right) > 0 \quad , \quad \beta_2\left(\frac{\omega_2^2}{4} - \lambda_0\right) < 0;$$

Therefore the solution (2.9a) in the neighbourhood of ω_2 is stable and is unstable close to ω_1. The result is shown in Fig. 2.2a where the unstable branches are denoted by broken lines.

2.2. The 2T-subharmonic resonance, its local stability and further period-doubling bifurcations

At frequencies ω far from ω_1 and ω_2 the assumption that C_0, C_1 are constant is too rough, and now we seek the 2T periodic solution in the whole range $\omega_1 - \omega_2$ in the form:

$$x_0^{(2)}(t) = A_0 + A_{1/2}\cos(\omega/2t + \phi) + A_1\cos\omega t, \tag{2.10a}$$

where $A_0, A_{1/2}, \phi, A_1$ need to be determined.

On inserting eqs. (2.10a) into eqs. (2.3) and applying the harmonic balance principle we obtain:

$$
\begin{aligned}
&-\frac{\omega^2}{4} + \frac{3}{4} A_{1/2}^2 + 3A_0^2 + \frac{3}{2} A_1^2 + 3 A_0 A_1 \cos 2\phi = 0 , \\
&- A_1 \omega^2 + \frac{3}{4} A_1^3 + 3A_0^2 A_1 + \frac{3}{2} A_{1/2}^2 A_1 + \frac{3}{2} A_0 A_{1/2}^2 \cos 2\phi = f , \\
&-\frac{1}{2} \gamma\omega + 3 A_0 A_1 \sin 2\phi = 0 , \\
&A_0^3 + \frac{3}{2} A_0 A_{1/2}^2 + \frac{3}{2} A_0 A_1^2 + \frac{3}{4} A_{1/2}^2 A_1 \cos 2\phi = P_0 ;
\end{aligned}
\tag{2.10b}
$$

Results of numerical evaluation for two examples, which differ only by the value of P_0 are shown in Fig. 2.3 and 2.8. The character of the resonance curves $A_{1/2}(\omega)$, $A_0(\omega)$, $A_1(\omega)$ in the two cases does not show substantial difference. However at $P_0 = 0.020$ the behaviour of the system is regular, periodic or very close to periodic in the whole range of the frequencies, while at $P_0 = 0.045$ chaotic motion was observed.

First we focus attention at the question of local stability of the subharmonic solution (2.10a). From (2.7) we learned that the left branch of $A_{1/2}(\omega)$ close to ω_1 is unstable. To extend validity of the criterion (2.9b) to larger values of $A_{1/2}$ we may replace values of C_0, C_1 at $\omega = \omega_1$ by the values of C_0, C_1 varying with ω. Such an approach to finding the criterion allows us to state that the left branch of $A_{1/2}(\omega)$ at $P_0 = 0.020$ is unstable from ω_1 up to $\omega = 0.89$ and at $P_0 = 0.045$ up to $\omega = 0.93$.

To answer the question of local stability in the whole range between ω_1 and ω_2 we derive variational equations for the disturbancy δx :

$$
\begin{aligned}
&\delta x = x(t) - x_0^{(2)}(t) , \\
&\delta\ddot{x} + \gamma\delta\dot{x} + \delta x[\lambda_0^{(2)} + \lambda_{1/2c} \cos \frac{\omega}{2} t +
\end{aligned}
\tag{2.11a}
$$

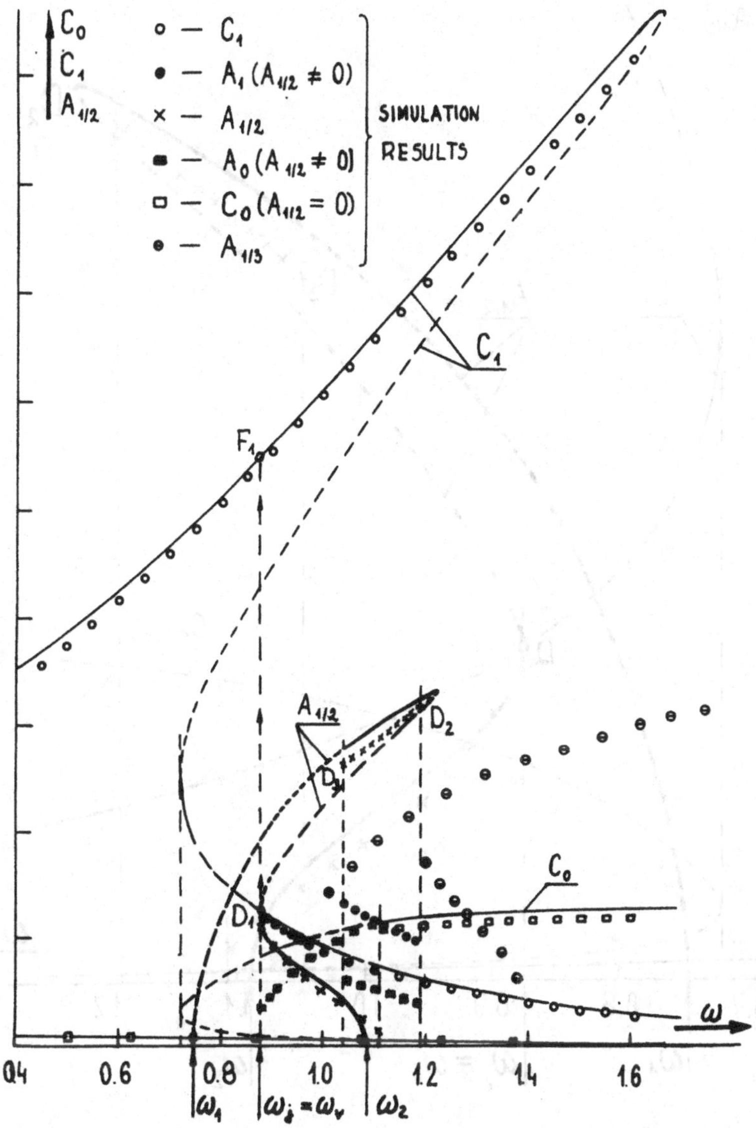

Fig. 2.3 (a). Resonance curves of the 2T-subharmonic solution; $P_0 = 0.020$, $f = 0.16$, $\gamma = 0.05$.

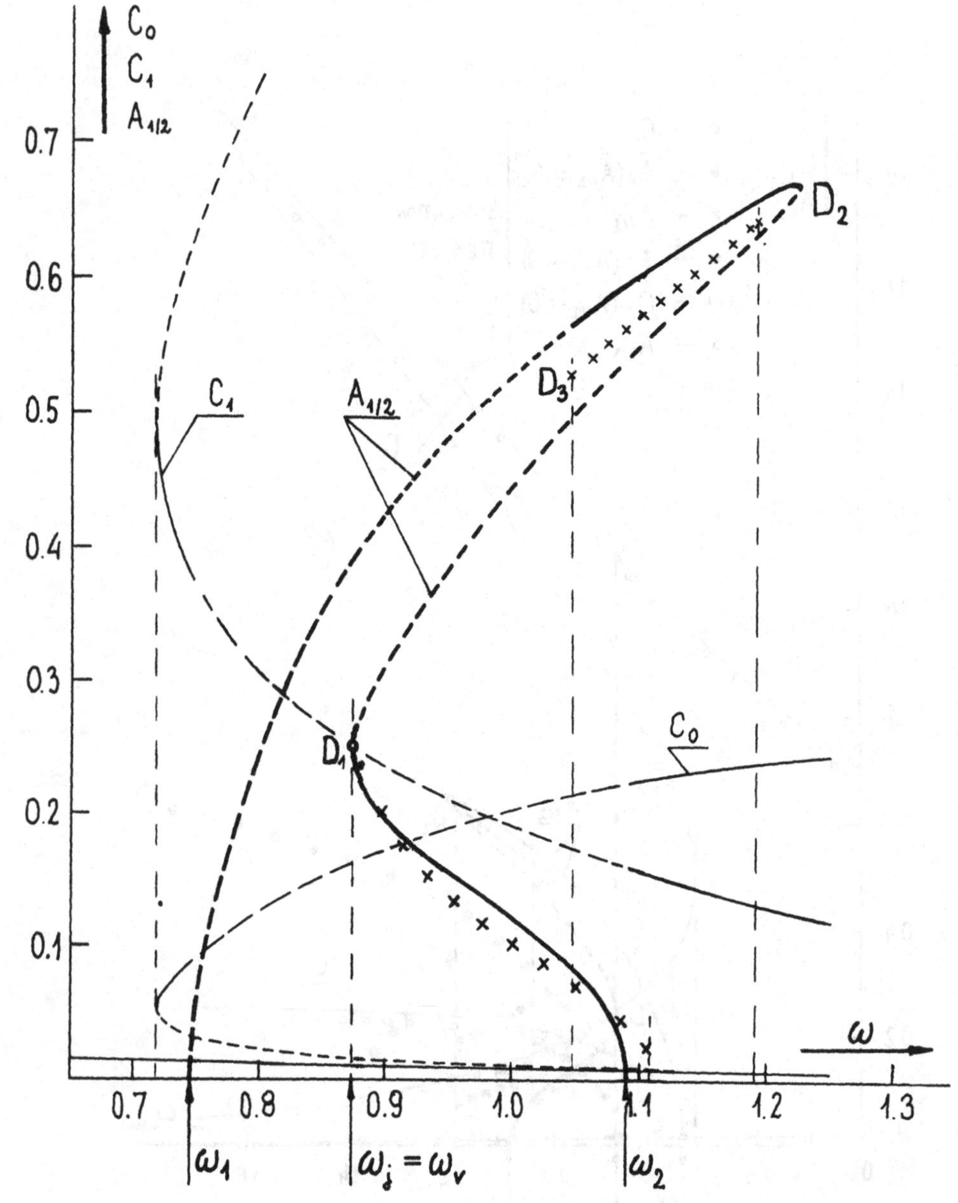

Fig. 2.3 (b). Resonance curves of the 2T-subharmonic solution; $P_0 = 0.020$, $f = 0.16$, $\gamma = 0.05$.

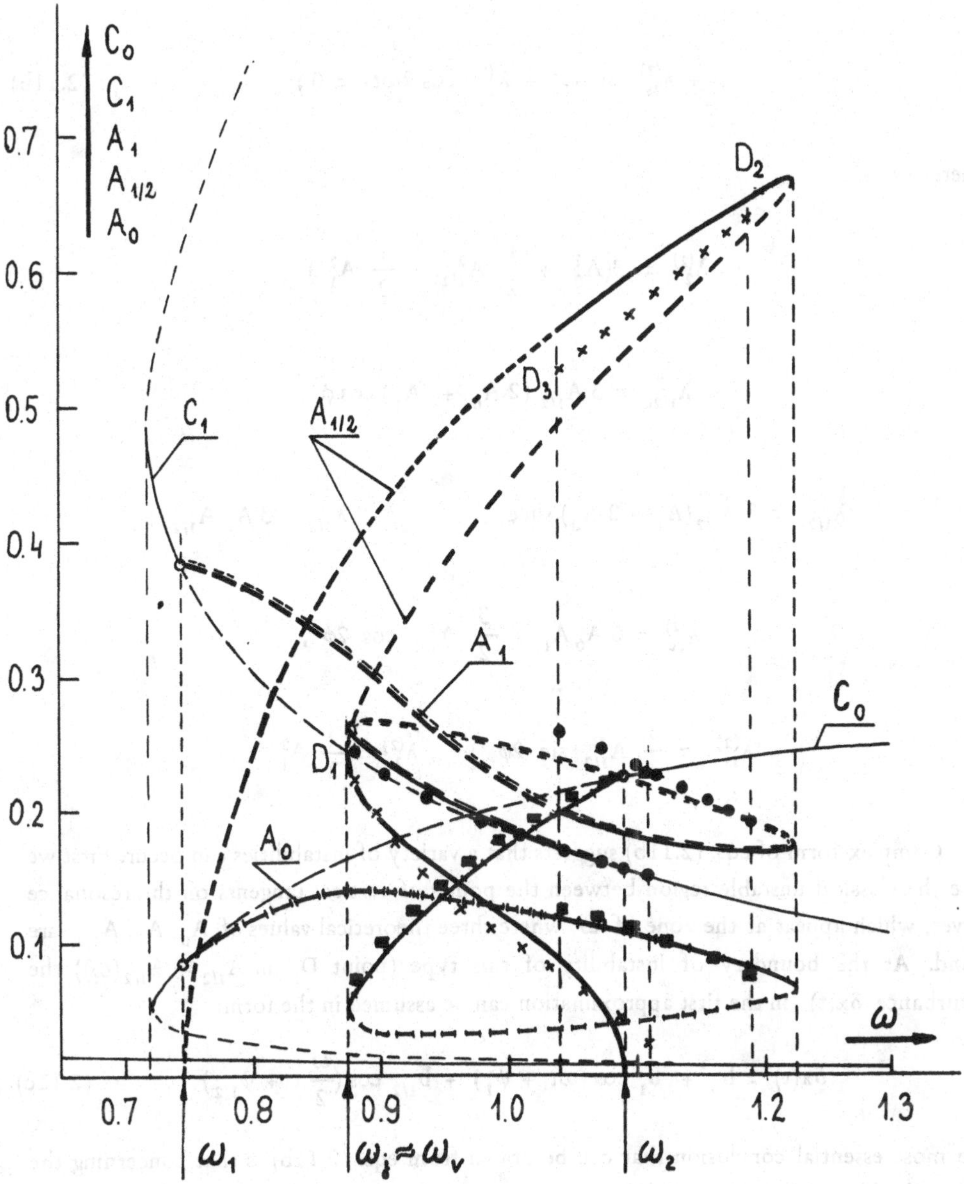

Fig. 2.3 (c). Resonance curves of the 2T-subharmonic solution; $P_0 = 0.020$, $f = 0.16$, $\gamma = 0.05$.

$$+ \lambda_{1/2s} \sin \frac{\omega}{2} t + \lambda_{3/2} \cos (\frac{3}{2} \omega t + \phi) + \lambda_{1c}^{(2)} \cos \omega t +$$

$$+ \lambda_{1s}^{(2)} \sin \omega t + \lambda_{2}^{(2)} \cos 2\omega t] = 0 ; \tag{2.11b}$$

where

$$\lambda_{2}^{(2)} = 3(A_0^2 + \frac{1}{2} A_{1/2}^2 + \frac{1}{2} A_1^2) ,$$

$$\lambda_{1/2c} = 3 A_{1/2} (2 A_0 + A_1) \cos \phi ,$$

$$\lambda_{1/2s} = 3 A_{1/2}(A_1 - 2 A_0) \sin \phi , \qquad \lambda_{3/2} = 3 A_1 A_{1/2} ;$$

$$\lambda_{1c}^{(2)} = 6 A_0 A_1 + \frac{3}{2} A_{1/2}^2 \cos 2\phi ,$$

$$\lambda_{1s}^{(2)} = \frac{3}{2} A_{1/2}^2 \sin 2\phi ; \qquad \lambda_{2}^{(2)} = \frac{3}{2} A_1^2 ;$$

Complex form of eqs. (2.11b) suggests that a variety of instabilities can occur. First we note the classical unstable region between the points of vertical tangents on the resonance curves, which appear at the zone of ω where three theoretical values of $A_0, A_1, A_{1/2}$ are found. At the boundary of instability of this type (point D_1 in $A_{1/2} \equiv A_{1/2}(\omega)$) the disturbance $\delta x(t)$ in the first approximation can be assumed in the form:

$$\delta x(t) = b_0 + b_1 \cos (\omega t + \phi_1) + b_{1/2} \cos (\frac{\omega}{2} t + \phi_{1/2}) ; \tag{2.11c}$$

The most essential conclusion that can be drawn from eqs. (2.11b) is that concerning the period doubling bifurcation. We see that a parametric term of the period of the solution under consideration appears as in eqs. (2.5d); now it is the term of the frequency $\omega/2$ and $3/2\ \omega$. It comes out that the lowest order unstable region is that associated with the terms $\lambda_{1/2}$, $\lambda_{3/2}$ and at the stability limit of the type we may seek an approximate solution as:

$$\delta x(t) = b_{1/4} \cos\left(\frac{\omega}{4} t + \phi_{1/4}\right) + b_{3/4} \cos\left(\frac{3}{4} \omega t + \phi_{3/4}\right) ; \qquad (2.11d)$$

Consequently a build-up of the $\omega/4$ and $3/4\,\omega$ harmonic components can be expected and the stable steady-state solution within the unstable zone may be sought in the form:

$$x_0^{(3)}(t) = A_0 + A_1 \cos \omega t + A_{1/2} \cos\left(\frac{\omega}{2} t + \phi_{1/2}\right) + A_{1/4} \cos\left(\frac{\omega}{4} t + \phi_{1/4}\right) +$$

$$+ A_{3/4} \cos\left(\frac{3}{4} \omega t + \phi_{3/4}\right) ;$$

Therefore we conclude that eqs. (211b) shows a possibility of further period doubling bifurcation – from 2T periodic solution to 4T periodic solution. In further steps of analogous analysis we notice a possibility of a build-up of the components with frequencies $3/8\,\omega$ and $5/8\,\omega$ and then generally $n \mp 1/2n$ $n = 2,4,8,16,\ldots$.

Summing up, using the variational Hill's type equation, which involves parametric excitation component of the period of the investigated steady-state solution, suggests a possible cascade of period doubling bifurcations. The hitherto investigations indicate that such cascade may lead to chaotic behaviour [7, 17].

2.3. Computer simulation resulsts and an approximate model of chaotic motion

To verify the theoretical approximate predictions, the eqs. (2.3) was simulated on an analog and then on digital computer and the response of the system was analysed by means of various techniques. First the dominant harmonic components in regular motion were detected: constant term, fundamental and $1/2\,\omega$ harmonic components. The harmonic component of the frequency $1/3\,\omega$ was also observed at certain non-zero initial conditions, because the $1/3\,\omega$ subharmonic resonance associated with the cubic nonlinearity is also predicted by first approximate solution.

This is an isolated solution not considered here [8].

From Figs. 2.1 and 2.3a we readily see that indeed the first approximate T-periodic solution (2.4a) describes the behaviour of the system with good accuracy in the whole range of frequency ω, where the solution is locally stable, i.e., on the upper resonant branch of $C_1(\omega)$ and on the lower branch except the region $\omega_1 - \omega_2$. On approaching the region from $\omega > \omega_2$ we reach point ω_2, where the amplitude $A_{1/2}$ appears and grows gradually with further decrease of the frequency, until a point close to the vertical tangent on the resonance curve ω_v. Values of the amplitude $A_{1/2}$ and of the two other components A_0 and A_1 are close to the theoretical ones.

At $\omega_j \approx \omega_v$ a jump phenomena occurs and after a short period of transient motion the response becomes close to the harmonic one with a large resonant amplitude $C_1(\omega)$.

The response associated with the upper branch of $A_{1/2}(\omega)$ was generated on applying proper nonzero initial conditions. It comes out that the branch is stable in a relatively narrow zone of frequency between points D_2 and D_3. It becomes evident that the whole remaining left branch of $A_{1/2}(\omega)$ and respective A_0, A_1 are unstable. At point D_3 the 2T-subharmonic response "jumps down" into the $1/3\,\omega$ subharmonic one (see also Fig. 2.4). Indeed in the whole range $0 < \omega < \omega_j$ the only steady-state solution that was generated for any initial conditions was the resonant large amplitude harmonic solution (upper branch of $C_1(\omega)$).

While the simulation results displayed in Figs. 2.1 and 2.3a,b,c confirm that the harmonic components assumed in solutions (2.4a) and (2.10a) predominate over other harmonics, they do not answer the question of period doubling bifurcation. The question might have been answered by detecting lower period and evidently low amplitude harmonic components in x(t). Instead the system response was analysed on the phase plane and Poincare maps were recorded. The sampling time for Poincare map was $T = 2\pi/\omega$ so that number of points m marked on $x(nT) - \dot{x}(nT)$ plane indicates the period of response $T_r = mT$.

Results presented in Fig. 2.5, 2.6 and illustrated in Fig. 2.4 confirm an appearance of the sequence of period doubling bifurcations. Regions of frequency at which the period of response was $T_r = 2T, 4T, \ldots$ are denoted in Figs. 2.4 for the lower as well as for the upper branch of $A_{1/2}(\omega)$. Figs. 2.5 shows time history, phase portraits and Poincare maps of the 2T and 4T periodic solution. Figs. 2.6a shows more complex pictures of motions. However, at further decrease of ω the response again becomes a regular one, and just before point $\omega_v \approx \omega_j$, has again period 2T. Thus the jump phenomena illustrated in Fig. 2.6d presents a jump from 2T to T periodic solution without any effects of earlier period doubling bifurcations.

The period doubling bifurcations occur also on the 3T subharmonic resonance and this is displayed in Fig. 2.7 in connection with Fig. 2.4.

Then we study behaviour of the system at $P_0 = 0.045$ for which the chaotic motion was observed by Ueda first [27]. The results displayed in Figs. 2.8a,b do not show any peculiarities. It looks that again in a wide zone of ω the predicted theoretical harmonic components predominate in the response. Similarly in the case $P_0 = 0.020$ we notice a sequence of period doubling bifurcations. The zones with $T_r = 2T, 4T, \ldots$ are denoted in Fig. 2.8b. Now, however, on further decrease of the frequency, the response does not return to the regular 2T periodic one and gradually the picture on Poincare map becomes more complex, (see 2.9a,b). Then at $\omega \approx 1.04$ there is a "burst out" of the "strange attractor",

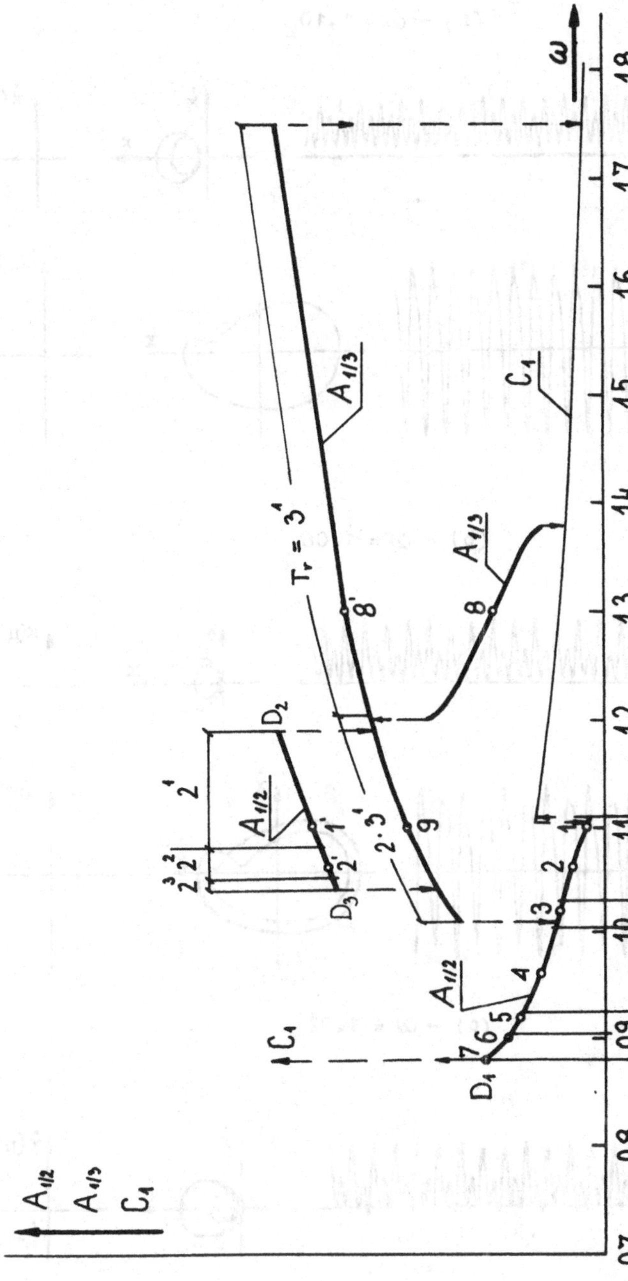

Fig. 2.4. Schematic diagram of bifurcation of the 2T and 3T-subharmonic resonances; $P_0 = 0.020$, $f = 0.16$, $\gamma = 0.05$.

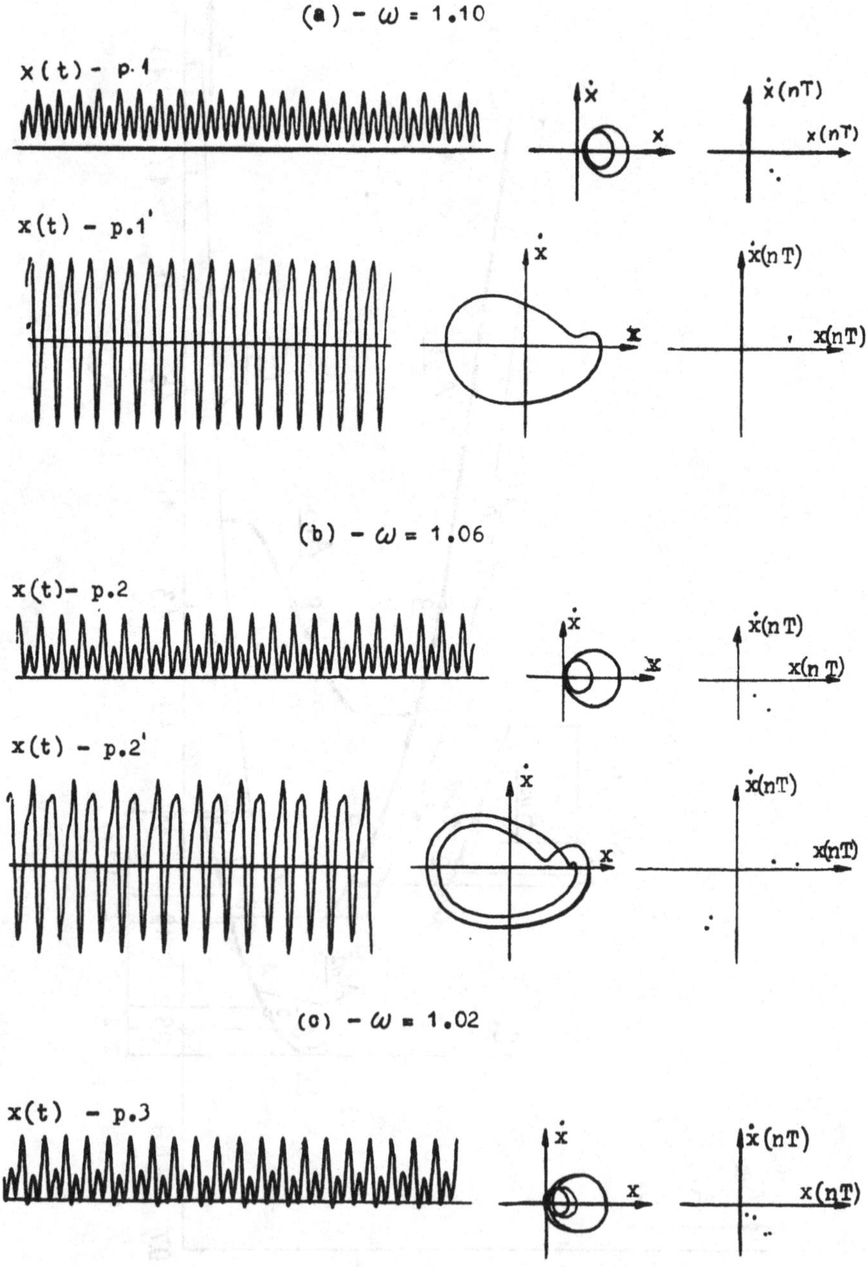

Fig. 2.5. Time history, phase portraits and Poincare maps at points denoted in Fig. 2.4.

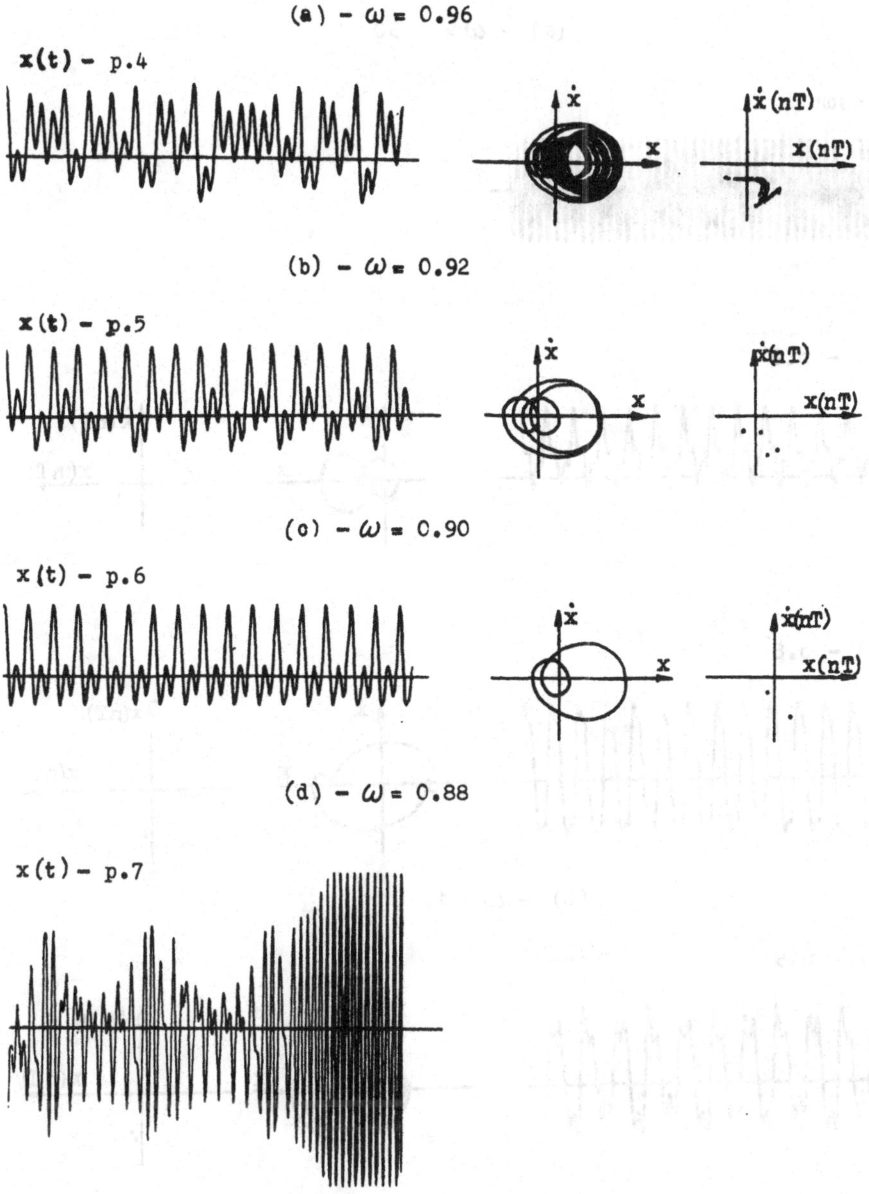

Fig. 2.6. Time history, phase portraits and Poincare maps at points denoted in Fig. 2.4.

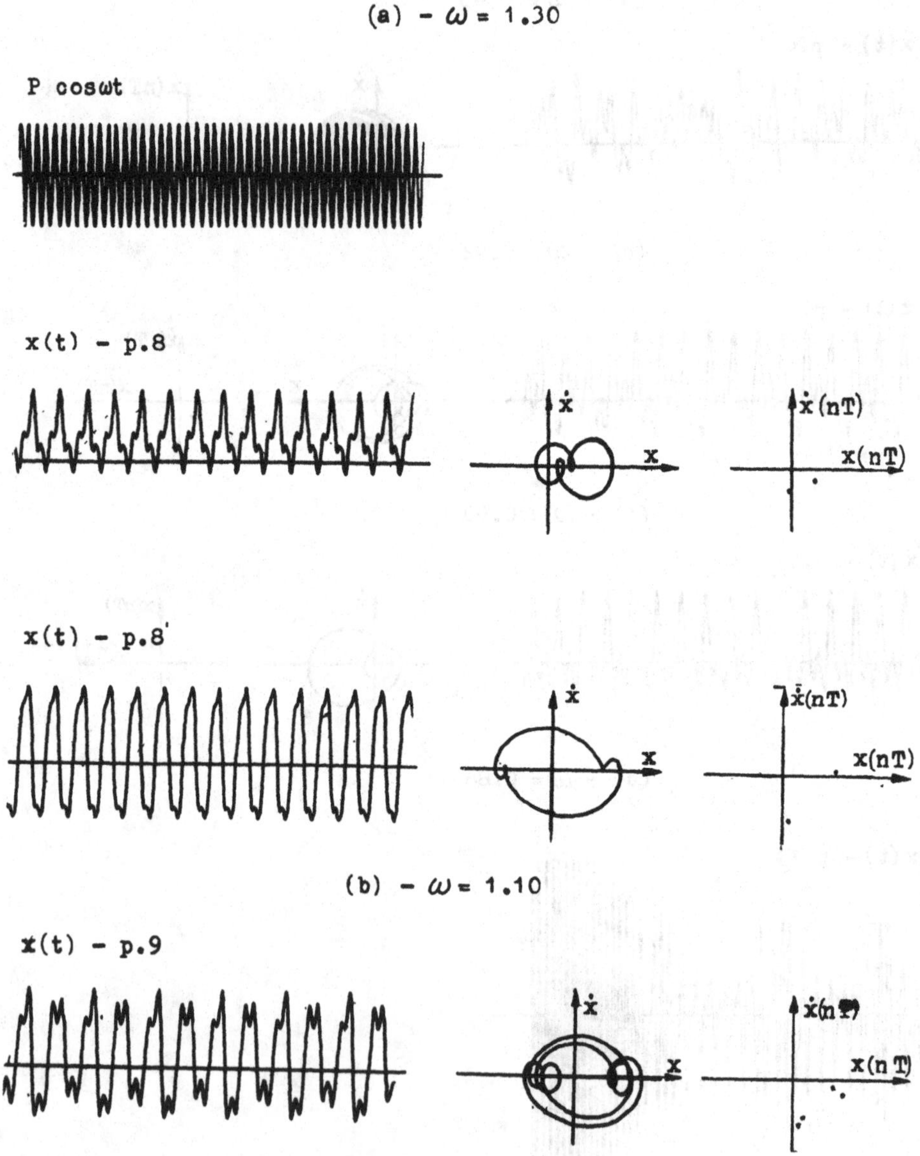

Fig. 2.7. Time history, phase portraits and Poincare maps of the 3T-subharmonic solution at points 8, 8′ and 9 in Fig. 2.4.

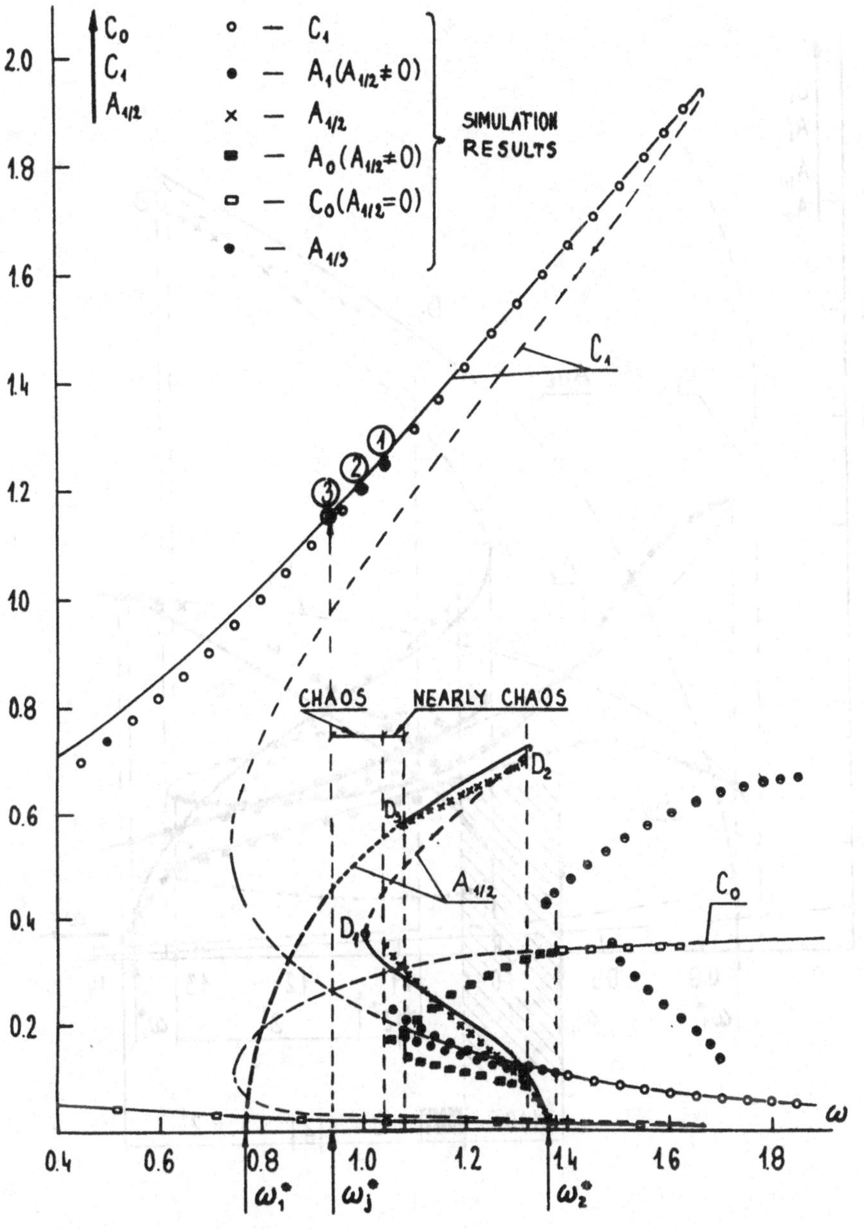

Fig. 2.8 (a). Resonance curves of the 2T-subharmonic solution; $P_0 = 0.045$, $f = 0.16$, $\gamma = 0.05$.

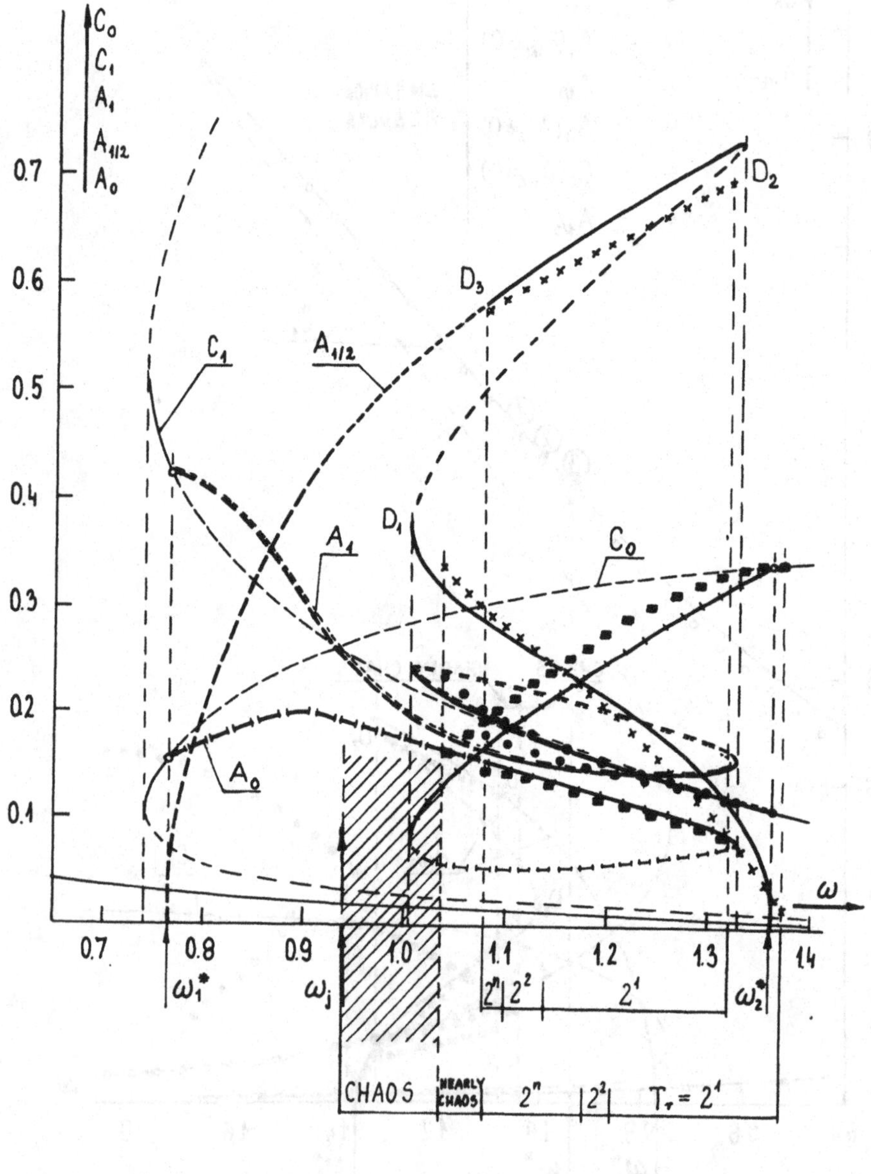

Fig. 2.8 (b). Resonance curves of the 2T-subharmonic solution, regions of periodic doubling bifurcation and chaotic behaviour.

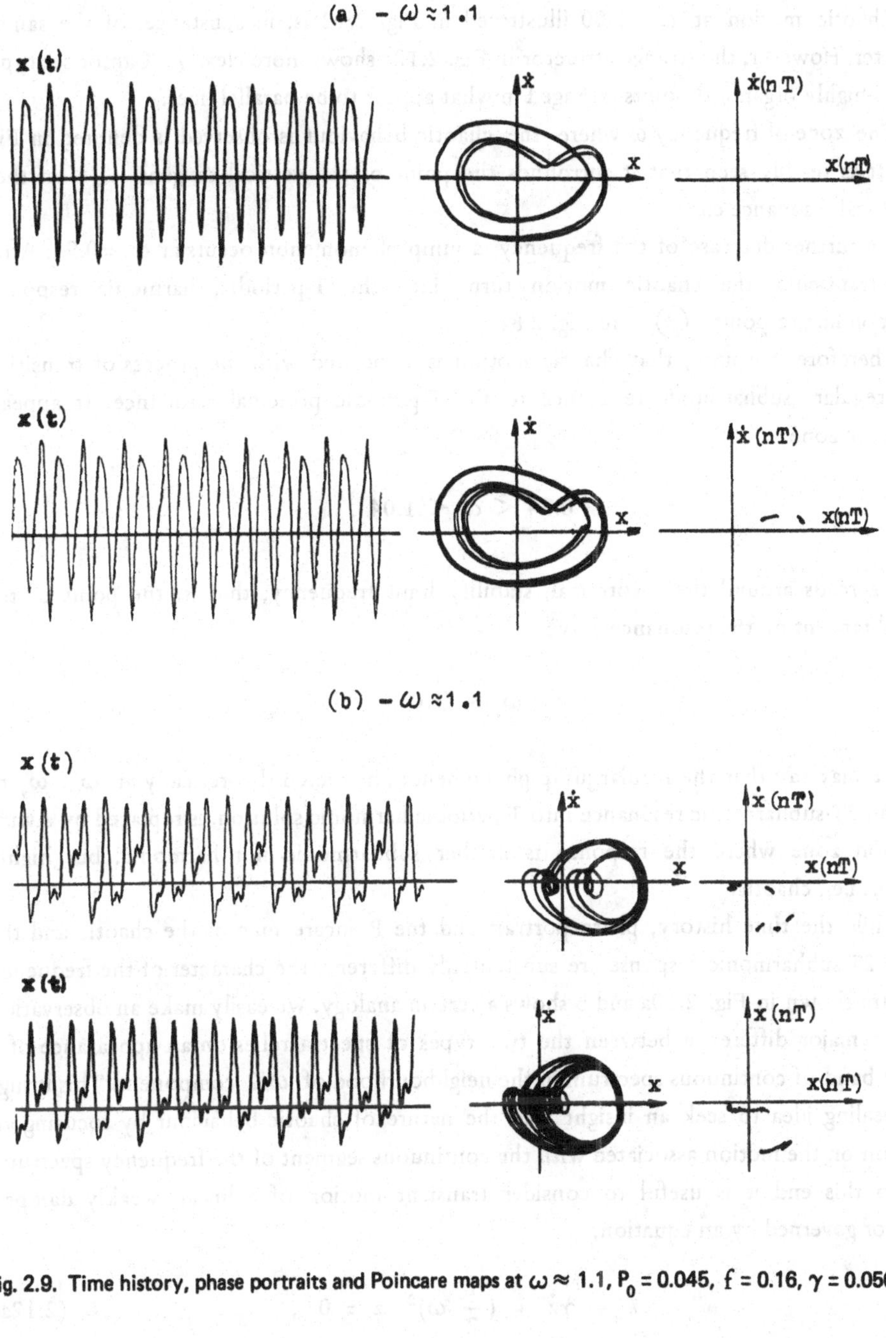

Fig. 2.9. Time history, phase portraits and Poincare maps at $\omega \approx 1.1$, $P_0 = 0.045$, $f = 0.16$, $\gamma = 0.050$.

with the results illustrated in Fig. 2.11. The appearance of continuous segments on the averaged power spectrum is, in addition to the characteristic picture on the Poincare map, strong evidence of chaotic motion (Fig. 2.10).

Chaotic motion at $\omega = 1.00$ illustrated in Fig. 2.12 is, in substance, of the same character. However, the strange attractor in Fig. 2.12c shows more clearly a Cantor set type pattern-highly organized points arranged in what appear to be parallel lines.

The zone of frequency ω where the chaotic behaviour is observed is denoted in Fig. 2.8b. It is readily seen that it surrounds the point of the vertical tangent ω_v on the theoretical resonance curves.

On further decrease of the frequency, a jump phenomenon occurs at $\omega_j = 0.94$. After some transients, the chaotic motion turns into the T-periodic, harmonic response, corresponding to point (3) in Fig. 2.8a.

Therefore, we note, that chaotic motion is associated with the process of transition from regular, subharmonic resonance to the T-periodic principal resonance. It appears within the zone

$$0.94 < \omega < 1.04 ,$$

which spreads around the theoretical, stability limit frequency, that is, the point of the vertical tangent on the resonance curve

$$\omega_v \approx 1.0$$

We may say that the regular jump phenomenon, predicted theoretically at $\omega = \omega_v$ to turn the 2T-subharmonic resonance into T-periodic harmonic solution, is replaced by a wide transition zone where the response is neither subharmonic nor harmonic, but highly irregular, i.e., chaotic.

While the time history, phase portrait and the Poincare map of the chaotic and the regular 2T-subharmonic response are substantially different, the character of the frequency spectrum shown in Fig. 2.10a and b shows a certain analogy. We easily make an observation that the major difference between the two types of spectra relies on an appearance of a narrow band of continuous spectrum in the neighbourhood of $\omega/2$ component. This brings an appealing idea to seek an insight into the nature of chaotic behaviour by focusing an attention on the motion associated with the continuous segment of the frequency spectrum.

To this end it is useful to consider transient motion of a linear, weakly damped oscillator governed by an equation,

$$\ddot{z} + \gamma\dot{z} + (\frac{1}{2}\omega)^2 z = 0 , \tag{2.12a}$$

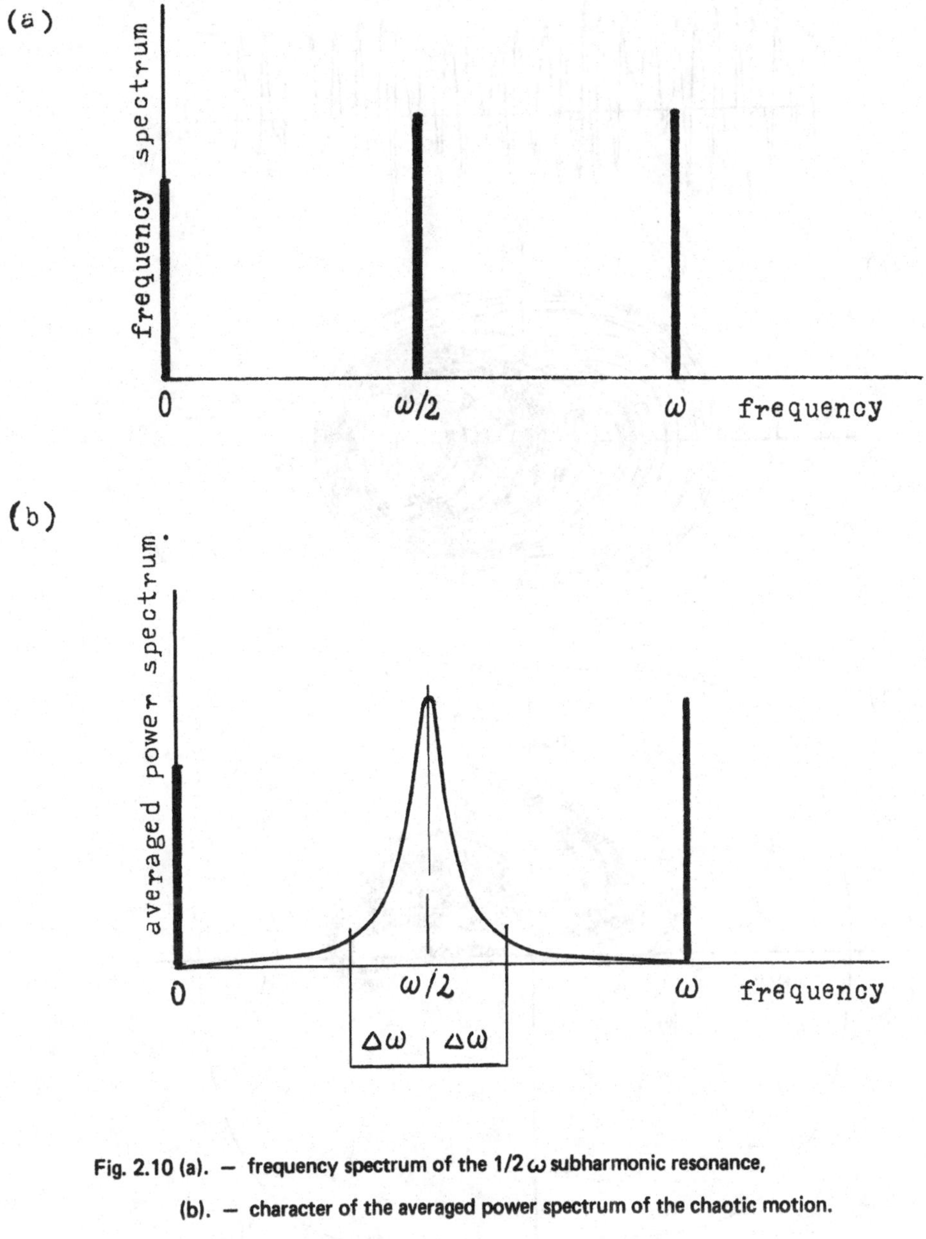

Fig. 2.10 (a). – frequency spectrum of the 1/2 ω subharmonic resonance,

(b). – character of the averaged power spectrum of the chaotic motion.

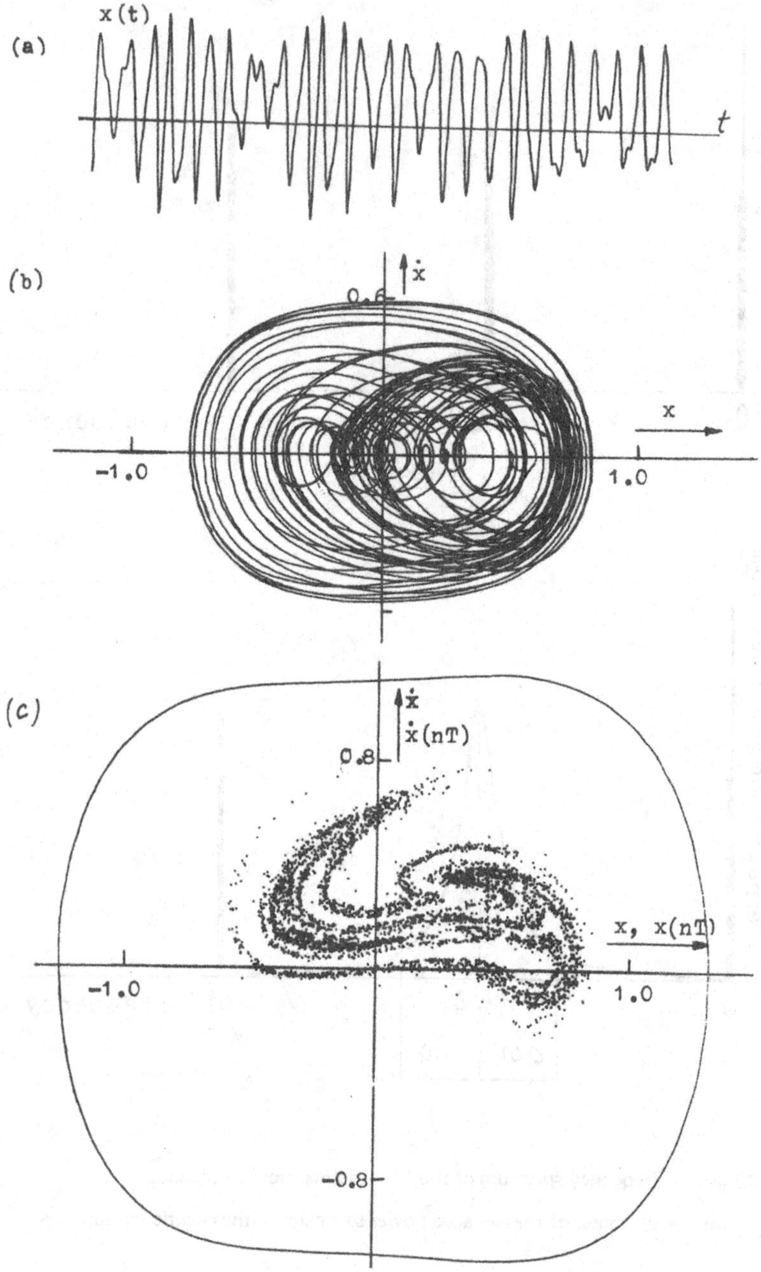

Fig. 2.11. Chaotic motion at $\omega = 1.04$; $P_0 = 0{,}045$; $f = 0{,}16$; $\gamma = 0{,}05$ - strange attractor. Full orbit in (c) illustrates the resonant harmonic solution-point ① in Fig. 2.8a.

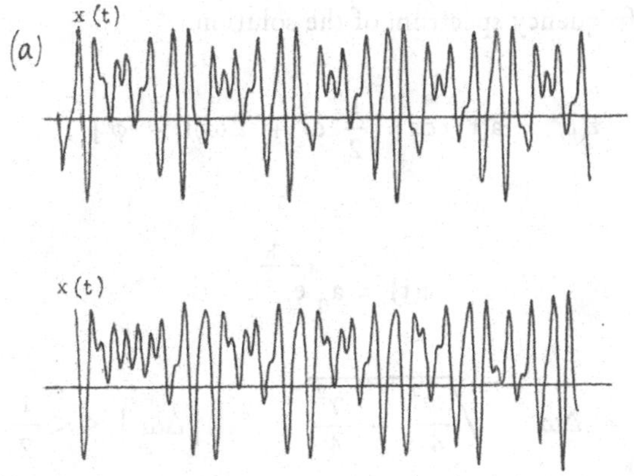

Fig. 2.12 (a). Time history of the chaotic motion at $\omega = 1.0$, $P_0 = 0.045$, $f = 0.16$, $\gamma = 0.050$.

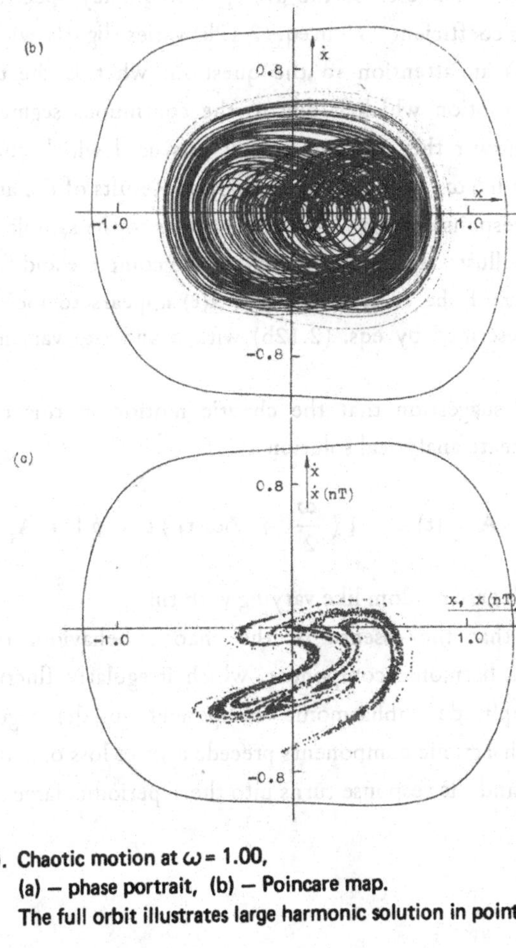

Fig. 2.12 (b,c). Chaotic motion at $\omega = 1.00$,
(a) – phase portrait, (b) – Poincare map.
The full orbit illustrates large harmonic solution in point ② in Fig. 2.8a.

and to notice that the frequency spectrum of the solution:

$$z(t) = a(t) \cos [(\frac{1}{2} \omega + \Delta\omega)t + \phi] , \qquad (2.12b)$$

where

$$a(t) = a_0 e^{-\frac{h}{2} t}$$

$$\frac{1}{2} \omega + \Delta\omega = \sqrt{\frac{\omega^2}{4} - \frac{\gamma^2}{4}} , \qquad |\Delta\omega| << \frac{1}{2} \omega$$

is very similar to that of the continuous segment of the chaotic motion [15]. Moreover it is essential to note that the character of the averaged frequency spectrum of z(t) remains unaltered if the damping coefficient γ in eq. (2.12b) varies slightly with time.

The analogy draws an attention to the question: what is the time history of the component of chaotic motion which is due to the continuous segment of the averaged power spectrum? To answer the question a filter was used which cut off the harmonic components higher than 0.8 ω in the chaotic response. Results of the analysis are shown in Fig. 2.13. The filtered response in Fig. 2.13a corresponds to the sample of chaotic response in Fig. 2.11a. Fig. 2.13b illustrates another sample of the complete and filtered time history. Indeed, the time history of the filtered response $\bar{x}(t)$ appears to look close to that of harmonic oscillations described by eqs. (2.12b) with a and $\Delta\omega$ varying randomly with time.

The result gives a suggestion that the chaotic motion in this case can be roughly described by an approximate analytical solution as

$$x(t) = A_0 + A_{1/2}(t) \cos [(\frac{\omega}{2} + \Delta\omega(t)) t + \phi] + A_1 \cos \omega t ,$$

where $A_{1/2}(t)$ and $\Delta\omega(t)$ are random-like varying with time.

Thus we can say that the essence of the chaotic behaviour relies on a sort of "instability" of the $\omega/2$ harmonic component, which irregularly fluctuates around the respective constant amplitude subharmonic component in the regular motion. The oscillations of the $1/2\omega$ harmonic components precede a strict loss of stability where the amplitude $A_{1/2}$ decays and the response turns into the T-periodic, large amplitude resonant solution.

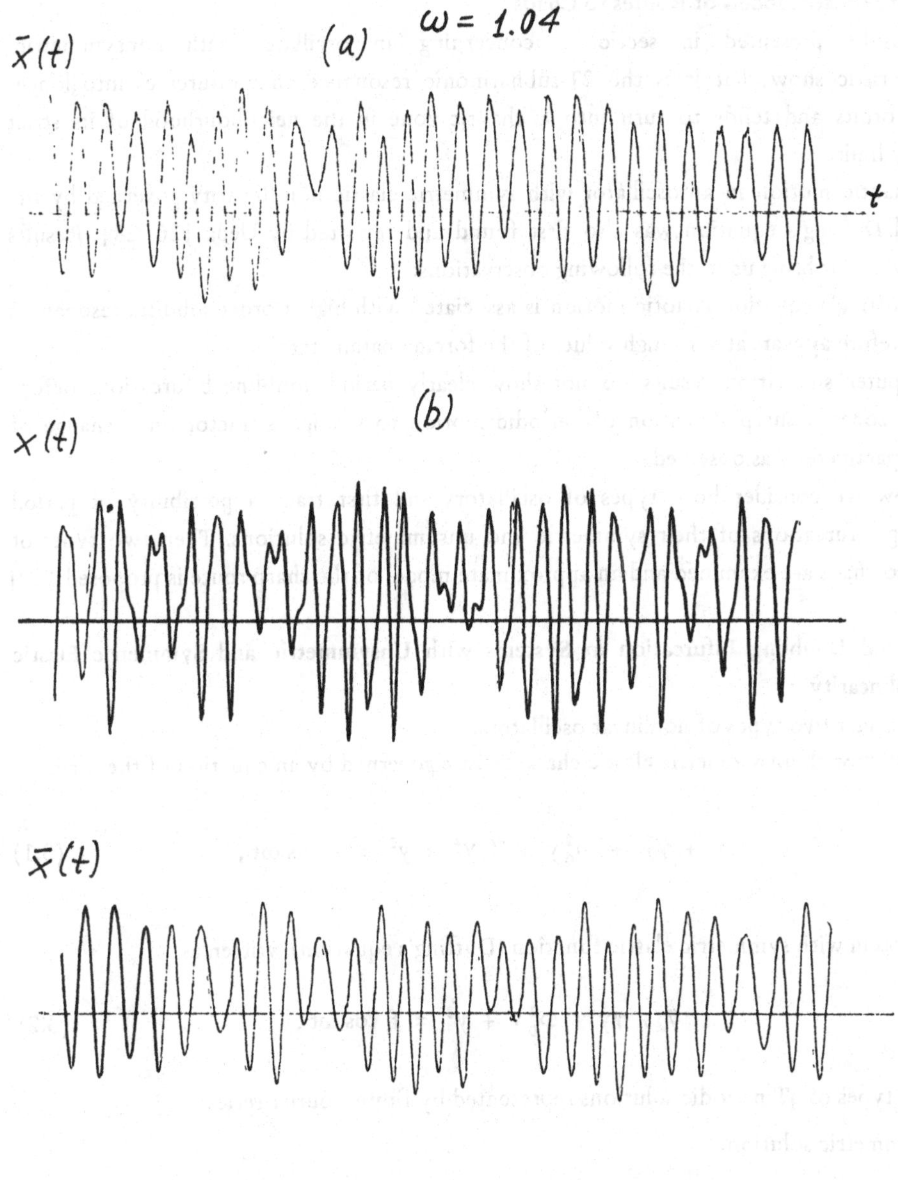

Fig. 2.13. Time history of chaotic and filtered responses at: $\omega = 1.04$, $P_0 = 0.045$, $f = 0.16$, $\gamma = 0.050$.

3. Oscillations with Unsymmetric and Symmetric Nonlinearity: Secondary Resonances and Approximate Models of Routes to Chaos

Results presented in section 2 concerning an oscillator with nonsymmetric characteristic show, that it is the 2T-subharmonic resonance, that bifurcates into longer period orbits and tends to turn into a chaotic zone in the neighbourhood of its strict stability limit.

Chaotic motion in an oscillator with symmetric elastic non-linearity, governed by the classical Duffing's equation, was also first found and reported by Ueda [26, 28]. Results obtained so far bring us to the following observations:

– in Duffing's equation chaotic motion is associated with higher order subultra resonances, and therefore appears at very high values of the forcing parameter;

– computer simulation results do not show clearly period doubling bifurcations before chaotic zone: a sharp transition of periodic motion to strange attractor on a change of system parameter was observed.

Now we consider both types of oscillators and first trace a possibility of period doubling bifurcations of their symmetric and unsymmetric solutions. Then two types of routes to chaos are examined and an approximate model of the sharp route is proposed [22].

3.1. Period Doubling Bifurcation in Systems with Unsymmetric and Symmetric Elastic Nonlinearity

Consider two types of nonlinear oscillators:

I – system with unsymmetric elastic characteristic governed by an equation of the form,

$$\ddot{y} + \gamma\dot{y} + \omega_0^2 y + \mathcal{K}_1 y^2 + y^3 = f \cos \omega t , \tag{3.1}$$

II – system with symmetric elastic function (Duffing's equation) written as

$$\ddot{y} + \gamma\dot{y} + \omega_0^2 y + y^3 = f \cos \omega t , \tag{3.2}$$

and two types of qT-periodic solutions represented by finite Fourier series:

an unsymmetric solution:

$$y_0^{(R)}(t) \neq - y_0^{(R)}\left(t + q\frac{T}{2}\right) = \sum_{p=1,2,3\ldots}^{R} A_p \cos\left(p\frac{\omega}{q}t + \vartheta_p\right) \tag{3.3}$$

and a symmetric solution:

$$y_0^{(R)}(t) = -\, y_0^{(R)}(t + q\frac{T}{2}) = \sum_{p=1,3,5}^{R} A_p \cos(p\frac{\omega}{q} t + \vartheta_p) \tag{3.4}$$

where

$$q = 1,2,3,\ldots \qquad , \qquad T = \frac{2\pi}{\omega} \quad ;$$

The solution (3.4) involves only odd harmonic components and is inherently associated with the symmetric system (3.2) The unsymmetric solution (3.3) containing a constant term and both even and odd harmonics is characteristic of the unsymmetric system (3.4). However, it may also appear in the system (3.2).

Since a cascade of period doubling bifurcations has been found in many physical systems as a very common route to chaos, it is essential to examine local stability of the solutions (3.3), (3.4) against a build-up of 2qT-periodic components.

Suppose that the coefficients A_p, ϑ_p, $p = 0,1,2,3,\ldots$ in equations (3.3), (3.4) have been determined by the harmonic balance method,and that one wishes to examine local stability of the qT-periodic solution thus evaluated. To this end one adds a small disturbance δy to the steady-state solution

$$\hat{y}(t) = y_0(t) + \delta y \ , \tag{3.5a}$$

inserts it into equation (3.1) and neglects terms of higher order in δy, with the result

$$\delta y + \gamma\delta\dot{y} + \delta y[\omega_0^2 + 2\mathcal{K}_1 y_0(t) + 3y_0^2(t)] = 0 \ ; \tag{3.5b}$$

On substituting the unsymmetric solution (3.3) one obtains the variational equaiton as

$$\delta\ddot{y} + \gamma\delta\dot{y} + \delta y[\omega_0^2 + \Pi_0 + \Pi_1(t) + \Pi_2(t)] = 0 \ , \tag{3.5c}$$

where $\Pi_1(t)$ is a periodic function of time with period of the solution $y_0(t)$:

$$\Pi_1(t) = \Pi_1(t + qT) \ , \tag{3.5d}$$

and $\Pi_2(t)$ denotes the harmonic components of period $1/2\ qT$:

$$\Pi_2(t) = \Pi_2(t + \frac{1}{2} qT) \ ; \tag{3.5e}$$

On applying Floquet theory (see e.g. [8, 9]) one can readily note that it is the term $\Pi_1(t)$, which is essential to the period doubling bifurcation. Only if

$$\Pi_1(t) \neq 0 , \tag{3.6a}$$

one may expect a particular solution in the form

$$\delta y(t) = e^{\epsilon t} \phi(t) \tag{3.6b}$$

where $\epsilon > 0$ and $\phi(t) = \phi(t + 2qT)$;

When this form of instability exists, a build-up of harmonic components of the period 2qT and consequently period doubling bifurcation becomes possible.

If $\quad \Pi_1(t) = 0 , \quad \Pi_2(t) \neq 0$

the only types of instability are those where the particular solution involves periodic function with period qT or 1/2 qT :

$$\delta(y(t) = e^{\epsilon t} \phi(t) ,$$

$$\phi(t) = \phi(t + qT) , \quad \text{or} \quad \phi(t) = \phi(t + \frac{1}{2} qT) \tag{3.7b}$$

(a) Unsymmetric System (3.1) and Unsymmetric Solution (3.3)

In this case the variational equation (3.5b) yields

$$\delta\ddot{y} + \gamma\delta y + \delta y[\omega_0^2 + \lambda_0 + \sum_{p=1,2,3\ldots}^{R} \lambda_{1p} \cos(p\frac{\omega}{q} t + \phi_{1p}) +$$
$$+ \sum_{p=1,2,3}^{R} \lambda_{2p} \cos(2p\frac{\omega}{q} t + \phi_{2p})] = 0 \tag{3.8a}$$

where

$$\lambda_0 \equiv \lambda_0(A_0, A_1, \ldots A_R , \quad \vartheta_1, \vartheta_2 \ldots \vartheta_R) ,$$

$$\lambda_{ip} \equiv \lambda_{ip}(A_0, A_1, \ldots A_R, \vartheta_1, \vartheta_2, \ldots \vartheta_R) ; \qquad i = 1,2.$$
$$p = 1,2, \ldots R.$$

It is worth noticing that

$$\lambda_{1p} = 0 \quad \text{only if} \quad \mathcal{K}_1 = 0 \quad \text{and} \quad A_0 = 0 ; \tag{3.8b}$$

It follows that period doubling bifurcations are very likely in the system with unsymmetric elastic function. It was shown in sec. 2 that it really happens even to the first approximate solution:

$$y_0^{(1)}(t) = A_0 + A_1 \cos(\omega t + \vartheta_1), \tag{3.9a}$$

when the variational equation (3.8a) becomes

$$\delta\ddot{y} + \gamma\delta\dot{y} + \delta y[\omega_0^2 + \lambda_0 + \lambda_1 \cos(\omega t + \vartheta_1) + \lambda_2 \cos(2\omega t + 2\vartheta_1)] = 0, \tag{3.9b}$$

where

$$\lambda_0 = \frac{3}{2} A_1^2 + 2\mathcal{K}_1 A_0 + 3A_0^2 ,$$

$$\lambda_1 = 2\mathcal{K}_1 A_1 + 6A_0 A_1 ; \qquad \lambda_2 = \frac{3}{2} A_1^2 ;$$

In equation (3.9b) the particular solution satisfying condition (3.6b) may exist and in the first approximation is sought as (compare to 2.6b)

$$\delta y(t) = e^{\varepsilon t} b_{1/2} \cos(\frac{\omega}{2} t + \phi_{1/2}) ; \tag{3.9c}$$

An unstable region of the type where $\varepsilon > 0$ may be found in the region of frequency

$$\omega \approx 2\sqrt{\omega_0^2 + \lambda_0} ; \tag{3.9d}$$

It follows that the $\omega/2$ harmonic component grows and consequently the 2T-subharmonic solution appears to be a stable steady-state solution in the region. Then the variatonal equation for a 2T-periodic approximate solution is assumed to have the form;

$$y_0^{(2)}(t) = A_0 + A_1 \cos(\omega t + \vartheta_1) + A_{1/2} \cos(\frac{\omega}{2} t + \vartheta_{1/2}) . \tag{3.9e}$$

This again shows a possibility of a period doubling bifurcation i.e. a possibility of the build-up of harmonic components of frequencies $\omega/4$ and $3/4\,\omega$, giving rise to 4T-periodic stable solution. In further steps a possibility of a cascade of period doublings can be seen.

(b) Symmetric System (3.2) and Symmetric Solution (3.4)

The qT periodic term in equation (3.5c) proves to vanish and the variational equation yields

$$\delta\ddot{y} + \gamma\delta\dot{y} + \delta y[\omega_0^2 + \lambda_0 + \sum_{p=1,3,5\ldots} \lambda_{2p} \cos(2p\frac{\omega}{q} t + \phi_{2p}) = 0, \qquad (3.10a)$$

where

$$\lambda_0 = \frac{3}{2} \sum_{p=1,3,5\ldots} A_p^2 ,$$

$$\lambda_{2p} \equiv \lambda_{2p}(A_1, A_3, A_5 \ldots A_R, \vartheta_1, \vartheta_3 \ldots \vartheta_R) ;$$

Particular solutions of eqs. (3.10a) are

$$\delta y(t) = e^{\epsilon t}\phi(t) ,$$

where

$$\phi(t) = \phi(t + \frac{1}{2} q T) , \quad \text{or} \quad \phi(t) = \phi(t + qT) ; \qquad (3.10b)$$

Consequently, in the unstable regions ony may expect a build-up of harmonic components with the frequency ω and its higher harmonics. Thus the analysis does not show a possibility of period doubling bifurcation. In spite of that a development of the symmetric solution into chaotic motion was observed and reported by Ueda (see [26] and [28]). The simulation results presented in the form of Poincare maps of the system response for a varying forcing parameter f at constant frequency $\omega = 1.0$ did not show period doublings before a burst-out of a strange attractor. This route to chaos, which appears to be a sharp one, will be considered in sec. 3.3.

(c) Symmetric System (3.2) and Unsymmetric Solution (3.3)

The unsymmetric solution does not exist in the symmetric system in the first approximation, i.e. at $R = 1$. Indeed A_0 vanishes identically at $R = 1$ and consequently the first approximate solution is

$$y_0^{(1)}(t) = A_1 \cos(\omega t + \vartheta_1)\,; \tag{3.11a}$$

Unsymmetric solutions do exist, however, in higher approximations. Indeed it was shown in earlier papers (see e.g. [20]) that at low frequencies and sufficiently high forcing parameter ψ there are regions where the harmonic solution is unstable and higher harmonic components become dominating ones. In the neighbourhood of the frequency

$$\omega \cong \frac{1}{2}\sqrt{\omega_0^2 + \lambda_0}\,, \tag{3.11b}$$

a stable solution is that which involves a constant term and the second harmonic component

$$y_0^{(2)}(t) = A_0 + A_1 \cos(\omega t + \vartheta_1) + A_2 \cos(2\omega t + \vartheta_2)\,; \tag{3.11c}$$

It is a general rule that the constant term is associated with even order harmonic components.

It follows that in the symmetric systems we may also obtain the variational equation in the form (3.8a).

Note that the coefficients λ_{1p} essential to the period doubling bifurcations vanish only when both $\mathcal{K}_1$ and A_0 vanish. In the case under consideration the coefficient $\mathcal{K}_1$ is equal to zero, however the constant term is not:

$$A_0 \neq 0\,;$$

Thus we conclude that periodic solutions in symmetric systems can also undergo period doubling bifurcation, provided they are the unsymmetrical ones.

3.2. Unsymmetric periodic solution and its transition to chaos via a cascade of period doubling bifurcations

It has been shown in sec. 2 and 3.1, that it is an unsymmetric periodic solution which can exhibit period doubling bifurcations whether the system is provided with symmetric or unsymmetric nonlinearity. Yet, since unsymmetric solutions in symmetric systems exist in higher approximations only, and hence appear at very high values of forcing parameter f, symmetric systems are not as ready as unsymmetric systems to show period doublings.

Now we reconsider the transition to chaotic motion via a cascade of period doublings in the unsymmetric system studied in sec. 2 and look for this route to chaos in the classical Duffing equation, i.e., in an example of a system with symmetric characteristics.

Bifurcation diagrams of the system considered in sec. 2 displayed in Fig. 3.1 are aimed

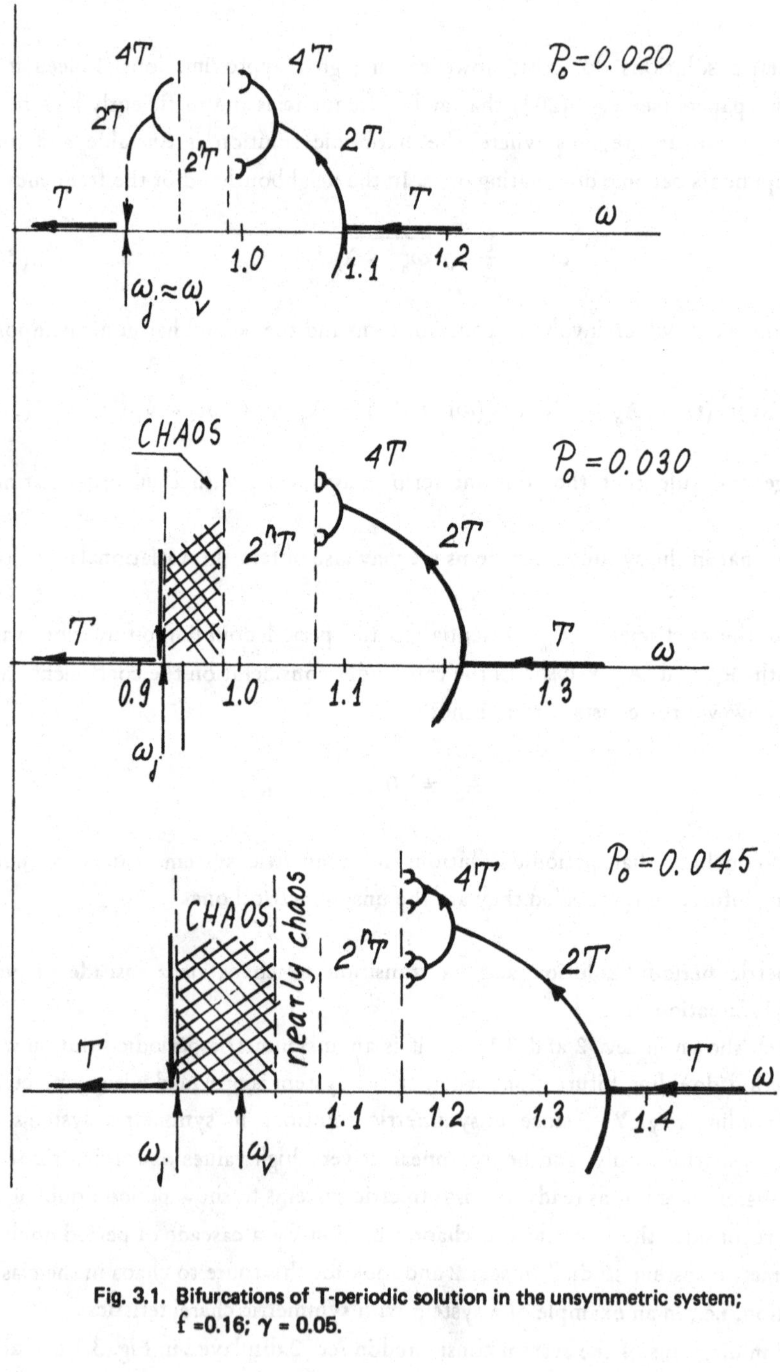

Fig. 3.1. Bifurcations of T-periodic solution in the unsymmetric system; f =0.16; γ = 0.05.

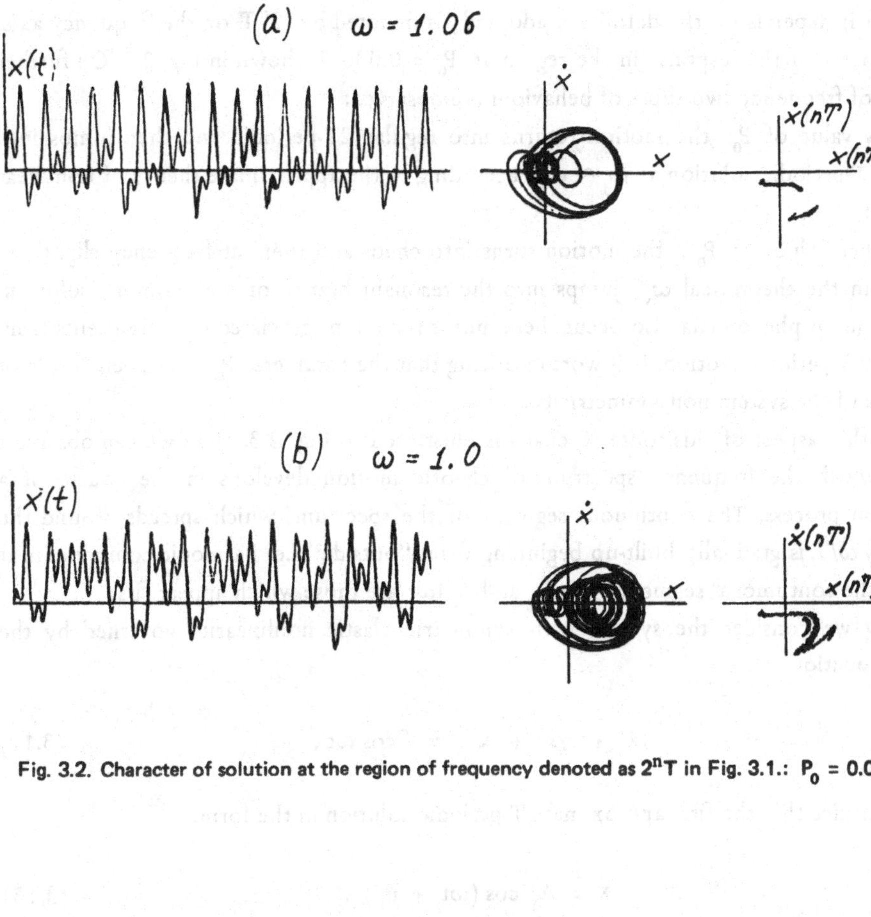

Fig. 3.2. Character of solution at the region of frequency denoted as 2^nT in Fig. 3.1.: P_0 = 0.030.

at illustrating schematically the system behaviour at three values of the parameter P_0. The T-periodic nonresontant solution,which appears at $\omega > \omega_2$, is denoted as T. The solution branches into 2T-periodic solution at $\omega = \omega_2$, and then 4T, 8T denote further bifurcations. A region when motion is no longer periodic but not completely chaotic, (although it depends on the definition adopted) is denoted by $2^n T$ on the frequency axis. The character of the response in the region at $P_0 = 0.030$ is shown in Fig. 3.2. On further decrease of frequency two types of behaviour are observed:

– at low value of P_0 the motion returns into regular 2T-periodic and then jumps into resonant T-periodic solution at $\omega_j \approx \omega_v$, according to the approximate theory of nonlinear vibration;

– at higher values of P_0 the motion turns into chaos and then, at frequency slightly lower than the theoretical ω_v, jumps into the resonant branch of the harmonic solution. Thus the jump phenomena also occurs here, but it is a jump associated with transients from chaotic to T-periodic motion. It is worth noticing that the parameter P_0 plays here a role of a measure of the system nonsymmetricity.

Another aspect of this route to chaos is illustrated in Fig. 3.3. Here we can observe a way in which the frequency spectrum of chaotic motion develops in the course of a bifurcation process. The continuous segment of the spectrum, which spreads around the frequency $\omega/2$, is gradually built-up beginning from "outside" i.e. harmonic components at edges of the continuous segment $1/4\omega$, and $3/4\omega$, are those which appear first.

Now we consider the system with symmetric elastic nonlinearity governed by the Duffing equation

$$\ddot{x} + \gamma\dot{x} + x^3 = f\cos\omega t\,, \tag{3.12}$$

and first notice that the first approximate T-periodic solution in the form:

$$x = A_1 \cos(\omega t + \vartheta_1)\,, \tag{3.13}$$

does not undergo period doubling bifurcations (see sec. 3.1c).

The lowest order solution, which can exhibit period doubling bifurcation, is an ultra-harmonic resonance of second order written as (see eqs. 3.11c):

$$x = A_0 + A_1 \cos(\omega t + \vartheta_1) + A_2 \cos(2\omega t + \vartheta_2) \tag{3.14}$$

which can arise in the neighbourhood of frequency $\omega \approx 1/2\ \omega(A_1)$.

One may view the result as a key to understanding why chaotic motion in Duffing equation is observed at very high values of the forcing parameter only.

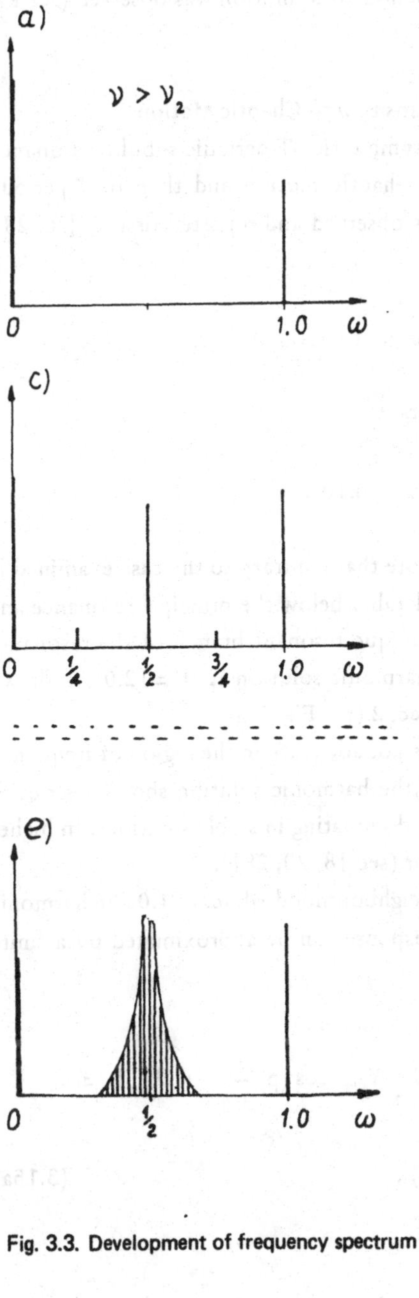

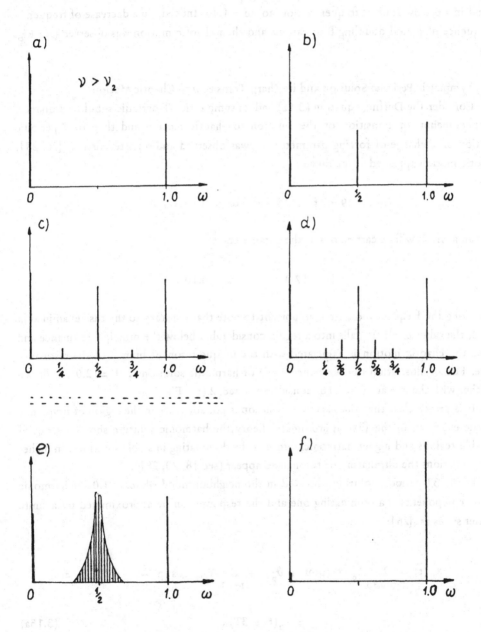

Fig. 3.3. Development of frequency spectrum in transition to chaos of the unsymmetric solution.

To verify theoretical predictions, that period doubling bifurcation may occur in the symmetric system, eqs. (3.12) was simulated on a computer, and the T-periodic unsymmetric solution close to that described by eqs. (3.14), was detected first. This type of response was found in a narrow zone of frequency close to $\omega = 1.40$. Indeed, on a decrease of frequency a sequence of period doubling bifurcations and then chaotic motion was observed (see Fig. 3.4).

3.3. Symmetric Periodic Solution and its Sharp Transition to Chaotic Motion

Consider the Duffing equation (3.12) and its symmetric 3T-periodic subultraresonance. A phenomenon of transition of the solution to chaotic motion and then to T-periodic solution on a change of forcing parameter f was observed and reported first in [26, 28]. Chaotic motion appeared at the range

$$9 < f < 13 \quad \text{at} \quad \omega = 1.0 \ .$$

Here an analysis will be carried out at the parameters

$$f = 12.0 \ ; \qquad \gamma = 0.10 \ ;$$

on varying the frequency ω. It is important to note that, contrary to the case examined in sec. 2, the value $\omega = 1.0$ falls into a region considerably below the principal resonance, and hence the chaotic motion is associated with the unique resonant branch of the resonance curve. For an illustration the resonance curve of harmonic solution at $f = 12.0$ is drawn together with the one at $f = 0.16$ considered in sec. 2 (see Fig. 3.5).

It is pretty clear that the harmonic solution is not adequate in the region of frequency considered. Even in the first approximate theory, the harmonic solution shows a series of unstable regions and higher harmonics prove to be dominating in stable solutions. In higher approximations the ultraharmonic resonances appear (see [8, 20, 28]).

In the 3T-periodic solution observed in the neighbourhood of $\omega = 1.0$, the harmonic $7/3\omega$ components is a dominating one and the response can be approximated by a finite Fourier series as [26]:

$$x_0(t) = \sum_{p=1,3,5} A_p \cos(p\omega t + \vartheta_p) + \sum_{p=1,7} A_{p/3} \cos\left(p\frac{\omega}{3} t + \vartheta_{p/3}\right) =$$

$$= x_0(t + 3T) \ ; \tag{3.15a}$$

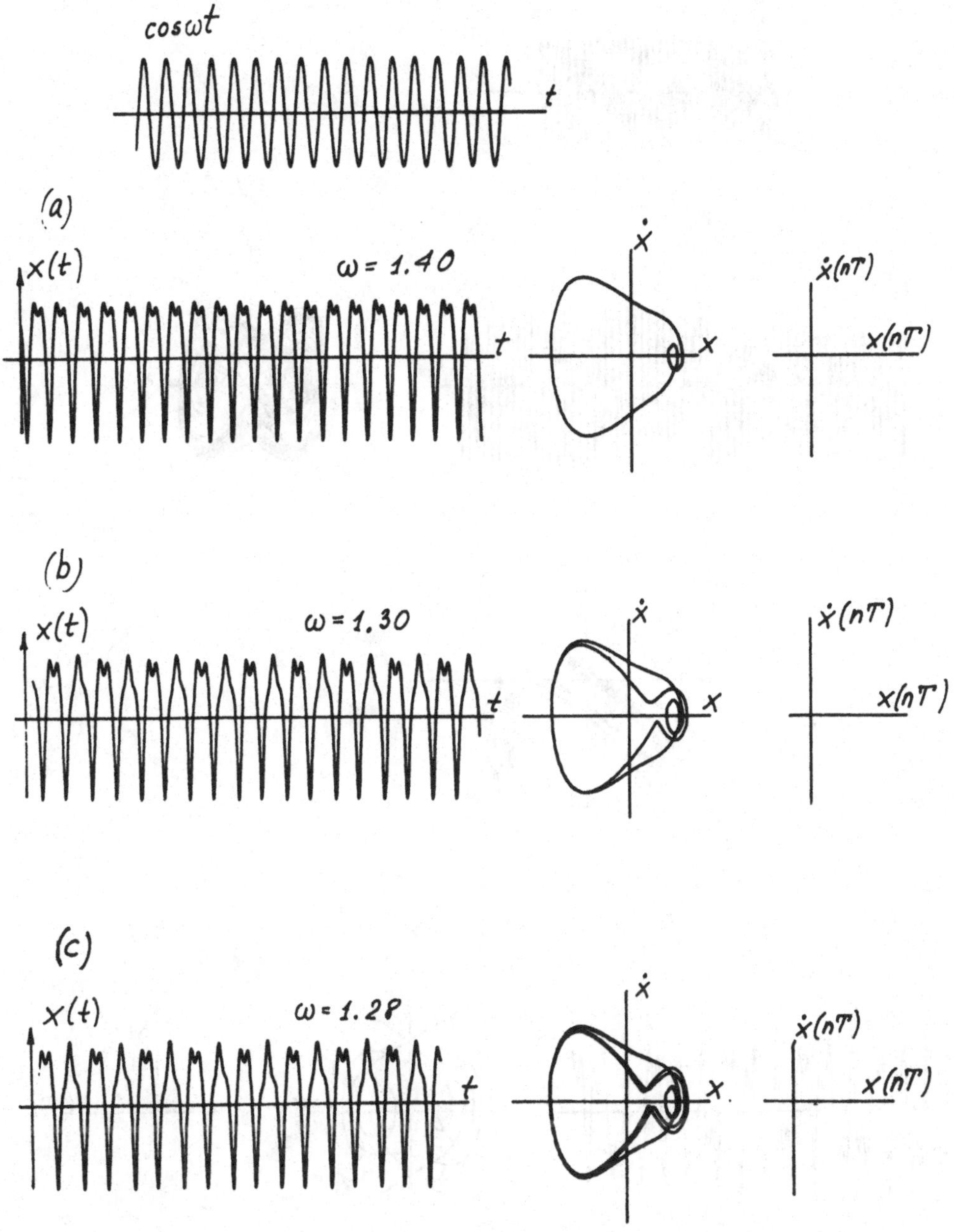

Fig. 3.4. Transition to chaos of unsymmetric solution in Duffing equation: a sequence of period doubling bifurcation.

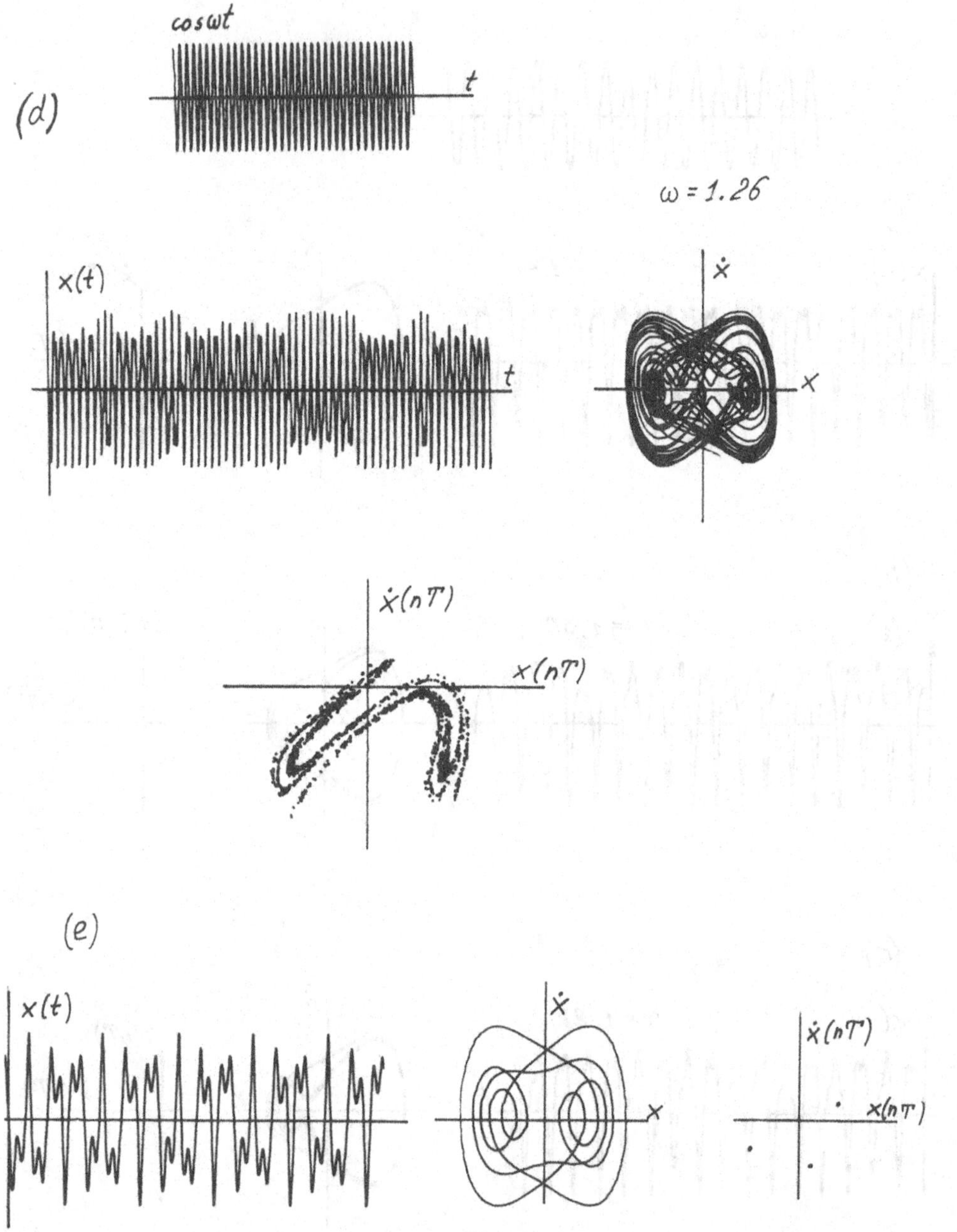

Fig. 3.4. continued

The first sum in equation (3.15a) involves the T-periodic components, and the second form – the 3T-periodic harmonics. The high number of harmonic components makes an analytical approach to an evaluation of resonance curves and stability limits unapplicable. General conclusion about the stability of the solutions presented in sec. 3.1 is, however, applicable: this is the case (b) with variational equation (3.10a) and consequently no period doubling bifurcation of the solution can be expected.

Yet,the solution does turn into chaotic motion upon a change of a parameter,and then to a T-periodic solution approximated by a series

$$x_{0_T}(t) = \sum_{p=1,3,5} A_p \cos(p\omega t + \vartheta_p) = x(t+T) ; \qquad T = \frac{2\pi}{\omega} ; \qquad (3.15b)$$

Although values of the amplitudes A_1, A_3, A_5 are certainly different than those in eqs. (3.15a), the essence of the difference between the two solutions lies in a disappearance of the 3T-periodic components.

At the transition zone, i.e. at the chaotic motion, continuous segments of the averaged power spectrum appear around the two 3T-periodic components (see Fig. 3.6). The zone of frequency, where the chaotic behaviour was observed, is marked in Fig. 3.5.

Further computer simulation results are displayed in Fig. 3.7a-e. First at $\omega = 1.06$ the regular 3T-periodic motion is recorded. Due to the large number of harmonic components the phase portrait looks very complex, however a closed orbit is clearly visible. Then at $\omega = 1.04$ we note that the motion is no longer strictly periodic: the phase portrait does not show a closed orbit and the three points on the Poincare map spread into what can be described as short arcs. On further decrease of the frequency a strange attractor appears rapidly and disappears also sharply terminating into the T-periodic solution (see Fig. 3.7cd,e).

It is clearly seen that the chaotic zone is not preceded by period doubling bifurcations. The only pre-chaotic motion is that shown in Fig. 3.7b. It is worth noticing that results obtained by Ueda [26] at $\omega = 10$ on changing the forcing parameter f showed very similar pre-chaotic behaviour.

What follows is an account of an attempt to propose an approximate model of the pre-chaotic motion, the motion characterized by three short segments of straight line around the three points of the Poincare map associated with the 3T-periodic solution. The starting point is the hypothesis presented in sec. 2. It says that the segments of continuous power spectrum surrounding sub-harmonic components can be interpreted as random-like fluctuations, or random-like transients of these harmonic terms. Since in the 3T-periodic solution (3.15a) it is the $7/3\omega$ harmonic component which is the dominating one, we can assume that in the pre-chaotic motion the amplitude $A_{7/3}$ begins to vary slightly with time oscillating with low frequency. This can be described as

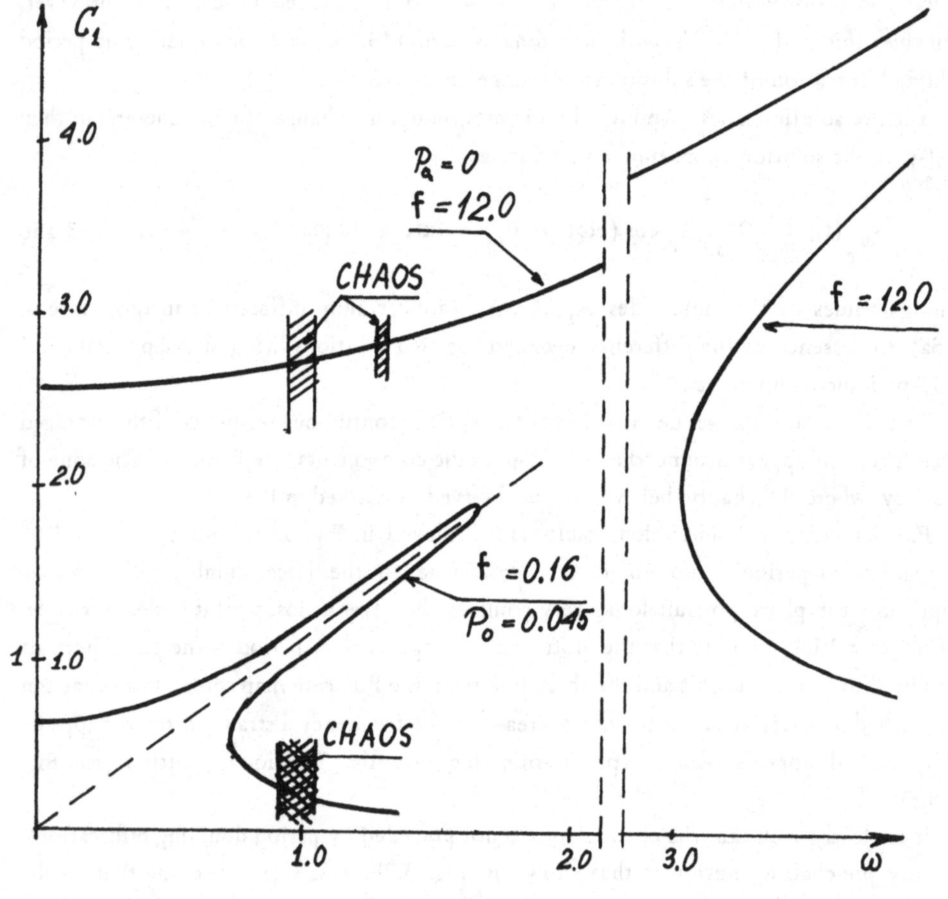

Fig. 3.5. Resonance curves of harmonic solution and zones of chaotic motion in symmetric and unsymmetric systems.

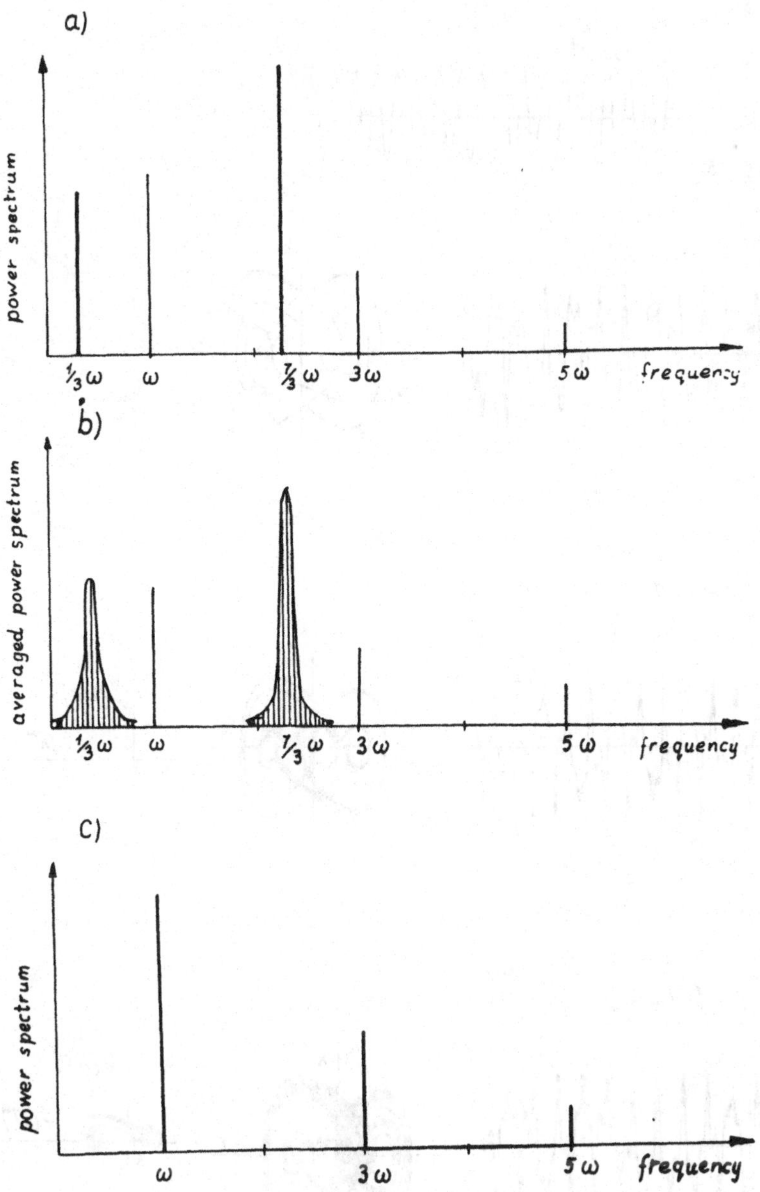

Fig. 3.6. Character of frequency spectrum of: (a) 3T-periodic motion; (b) chaotic motion; (c) T-periodic motion.
[(b) - Ueda [26]).

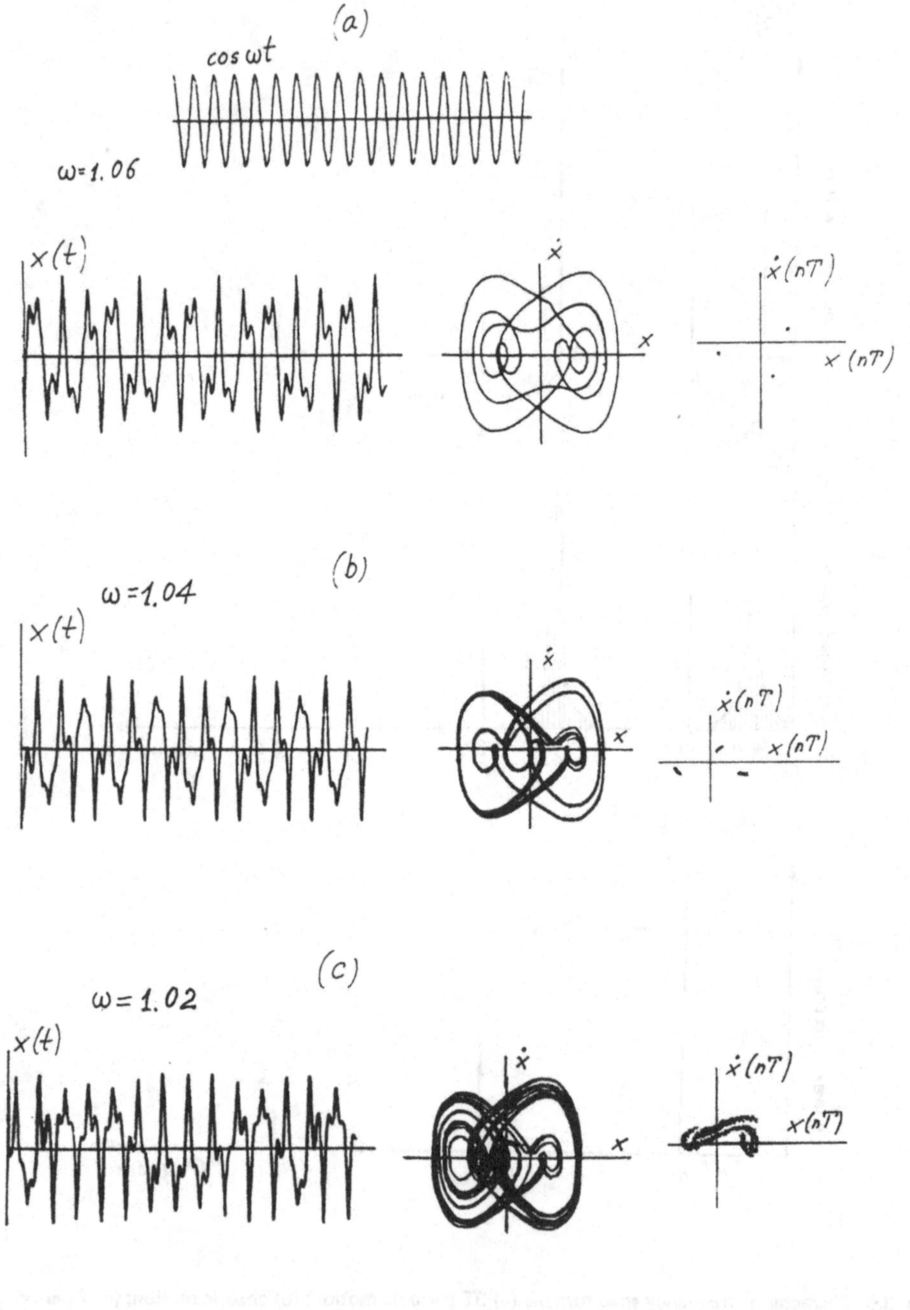

Fig. 3.7. Time histories, phase portraits and Poincare maps in sharp transition to chaos.

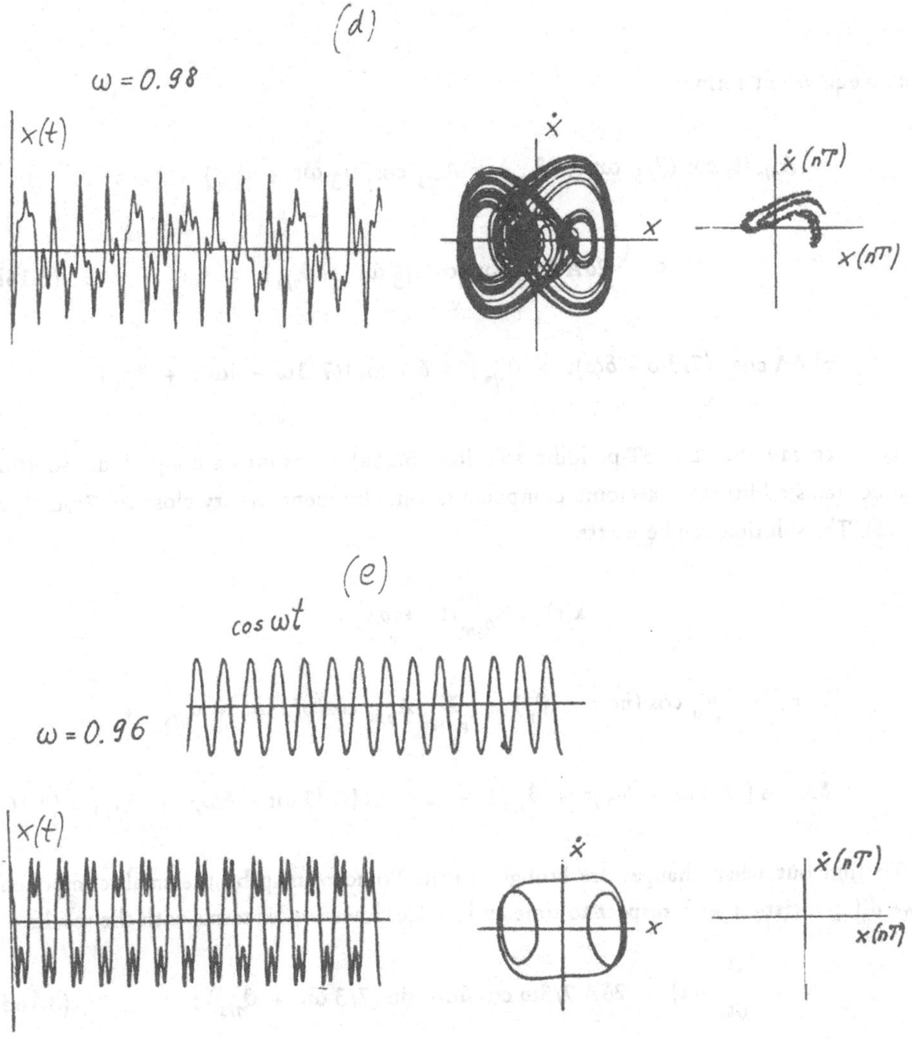

Fig. 3.7. continued

$$[A_{7/3} + 2\delta A \cos \delta\omega t] \cos (7/3\, \omega t + \vartheta_{7/3}) \,, \tag{3.16a}$$

or in an equivalent form as

$$A_{7/3}(t) \cos (7/3\; \omega t + \vartheta_{7/3}) = A_{7/3} \cos(7/3\; \omega t + \vartheta_{7/3}) + \delta x \;;$$

$$\delta x = 2\delta A \cos \delta\omega t \cos(7/3\; \omega t + \vartheta_{7/3}) = \tag{3.16b}$$

$$= \delta A \cos [(7/3\omega + \delta\omega)t + \vartheta_{7/3}] + \delta A \cos [(7/3\omega - \delta\omega)t + \vartheta_{7/3}] \;;$$

One can say that the 3T-periodic solution (3.15a) turns into a nonperiodic solution, which contains additional harmonic components with frequencies very close to $7/3\omega$ (see Fig. 3.8). The solution can be written as:

$$x(t) = x_{0_{3T}}(t) + \delta x =$$

$$= \sum_{p=1,3,5} A_p \cos (p\omega t + \vartheta_p) + \sum_{p=1,7} A_{p/3} \cos(p/3\, \omega t + \vartheta_{p/3}) +$$

$$+ \delta A \cos [(7/3\,\omega + \delta\omega)t + \vartheta_{7/3}] + \delta A \cos [(7/3\, \omega t - \delta\omega)t + \vartheta_{7/3}] \;; \tag{3.16c}$$

To find out what changes are brought to the Poincare map by the small component δx, we differentiate it with respect to time and neglect lower order terms with the results

$$\frac{d}{dt}(\delta x) \equiv 2\delta A\, 7/3\omega \cos \delta\omega t \,\sin (7/3\, \omega t + \vartheta_{7/3}) \,; \tag{3.16d}$$

By virtue of eqs. (3.16) the following observation can be made:

– if the 3T-periodic solution is mapped on the $x(nT) - \dot{x}(nT)$ plane by the three points with coordinates X_i, Y_i (see Fig. 3.9a), then the disturbance δx results, after sufficiently long time interval, in an appearance of large number of points within regions

$$X_i \mp \delta A \;; \qquad Y_i \mp \delta A\, 7/3\,\omega \;; \qquad i = 1,2,3.$$

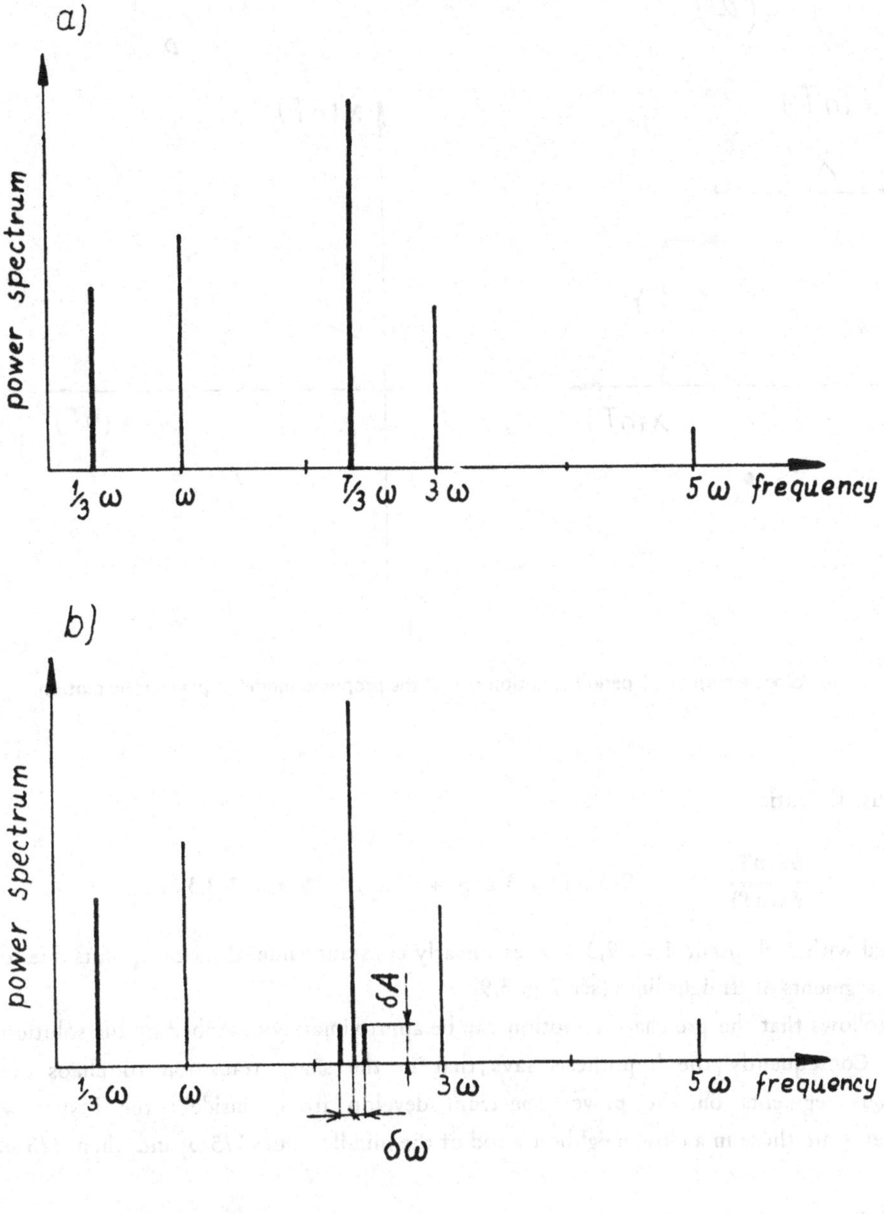

Fig. 3.8. Character of frequency spectrum of 3T-periodic solution and proposed model of pre-chaotic motion.

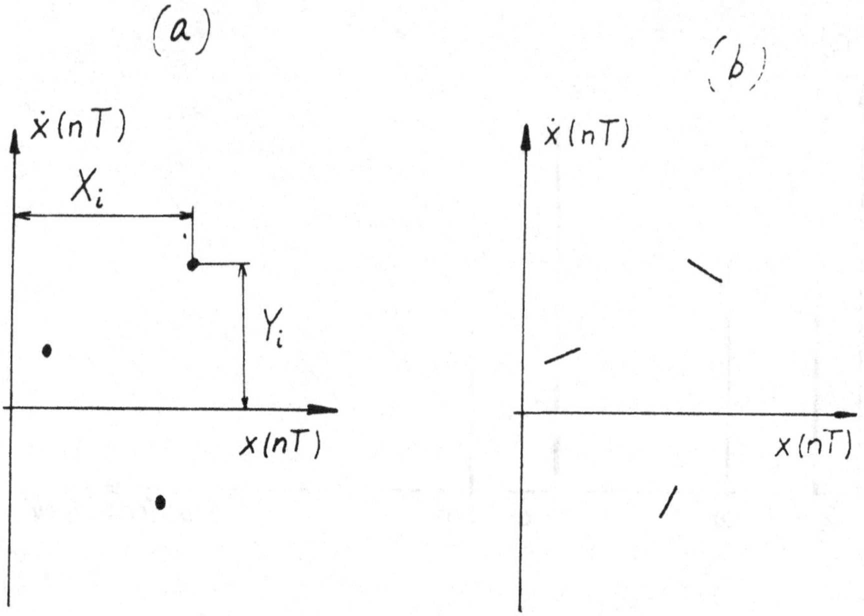

Fig. 3.9. Poincare map of 3T-periodic motion and of the proposed model of pre-chaotic motion.

– Because the ratio

$$\frac{\delta \dot{x}(nT)}{\delta x(nT)} \cong -\, 7/3\ \nu\ \mathrm{tg}(7/3\ 2\pi n + \vartheta_{7/3})\,, \qquad n = 1,2,3\ldots$$

associated with each point $i = 1,2,3$ takes a nearly constant value, the set of points due to δx form segments of straight lines (see Fig. 3.9b).

It follows that the pre-chaotic motion can be approximately described by the solution (3.16c). Consequently, the hypothesis says, that in the sharp transition to chaos the continuous segments on the power spectrum develop from "inside": the first new components are those in a close neighbourhood of the middle values $7/3\omega$ and then $1/3\omega$.

Conclusions

A combination of computer simulation analysis and theoretical evaluations of approximate periodic solution for two nonlinear oscillators with single equilibrium positions leads to the observation that chaos is a transition zone between a qT-subharmonic solution and a T-periodic solution. In the theory of nonlinear oscillations the transition is due to occur at points of vertical tangent on the resonance curves of the qT-periodic solution. The observations presented indicate that when a system parameter exceeds a certain critical value, the region qT-periodic motion is separated from the region of T-periodic motion by a zone of chaotic behaviour. The character of the averaged power spectrum associated with the chaotic motion gives the appealing idea of interpreting the motion as random-like fluctuations of the qT-periodic harmonic components i.e. the components which are losing stability and then decay in the region of T-periodic motion.

Two routes of chaos are discussed and illustrated:

– the classical route to chaos via a cascade of period doubling bifurcations. This happens to unsymmetric solutions and hence is associated with systems having unsymmetric nonlinear characteristic. In the route to chaos the continuous segments of the averaged power spectrum are gradually built-up from "outside" i.e. components on the edges of a segment appear first.

– the "sharp" route to chaos is associated with symmetric periodic solutions and hence the system with symmetric nonlinear characteristic. On studying the pre-chaotic behaviour an approximate mathematical model of the route to chaos is proposed. The hypothesis states that the continuous segments of the averaged power spectrum begin to develop from "inside" i.e. components in a close neighbourhood of the middle frequencies appear first.

It has been shown that symmetric solutions, the first approximate harmonic solution in the Duffing equation in particular, do not undergo period doubling bifurcations, and the conclusion has been drawn that symmetric systems are not ready to exhibit chaotic motion. This is viewed as a key to understanding why chaotic motion in the Duffing equation occurs at very high values of forcing parameter. At the high values of f periodic solutions involve a large number of harmonic components and consequently chaotic zones are situated beyond the system parameters treated by the approximate analytical methods.

4. An Oscillator with Three Equilibrium Positions. An Approximate Criterion for Chaotic Motion

The phenomena of chaotic motion in a system having three positions of equilibrium governed by an equation

$$\ddot{x} + \gamma\dot{x} - \frac{1}{2}(1 - x^2)x = f \cos \omega t ,$$

has attracted a great deal of attention in the recent literature. The system was studied first by Holmes [2, 3] and has become now a classical mathematical model of a buckled beam and a model of systems which exhibit chaotic motion. A main point of interest is a "strange behaviour" of the system: when the ratio f/γ exceeds a certain critical value, periodic oscillations around one of the two stable equilibrium positions turn into an irregular motion consisting essentially in random-like jumps from oscillations around one to oscillations around the other rest point. The phenomenon was first observed experimentally by Tseng and Dugundji as early as in 1971 [24]. However it did not draw much attention and was described as "snap through oscillations".

In a series of papers [4,5,6,13] major interest was focused on finding and recording essential characteristics of chaotic motion either with experimental or computer simulation methods, theoretical analysis of strange attractors by means of topological methods and routes to chaos. Some attempts to develop an approximate criterion for chaotic motion i.e. to determine critical parameters for which one might expect chaotic behaviour was made by Holmes and Moon [3,13]. However, theoretical and experimental results of boundaries between regular and chaotic motion in [13] do not show satisfactory coincidences.

In this section we first study the approximate harmonic solution of the system and then determine regions of regular and chaotic motion in the frequency-forcing parameter plane. An analysis of stability of Small Orbit harmonic solution makes it possible to propose an approximate criterion for the boundary of periodic motion and its transition to irregular behaviour. Although a computer simulation shows a variety of subharmonic motions, the harmonic solution with period of excitation appears to play an essential role in the task of determining critical values of system parameters for the chaotic motion to occur [16].

4.1. General Equation and its Fundamental Properties

To derive the second order differential equation whose solutions are to be studied we consider partial differential equation for transverse large deflections of a beam accompanied with linear autonomous boundary conditions written in general form as

$$m\frac{\partial^2 w}{\partial t^2} + H\frac{\partial w}{\partial t} + EJ\frac{\partial^4 w}{\partial y^4} + F_0\frac{\partial^2 w}{\partial y^4} + L(w) = p(y) \cos \bar{\nu}\tau ;$$

$$B_{i1}(w)_{y=0} = 0 \; ; \quad B_{i2}(w)_{y=\ell} = 0 \; ; \quad i = 1,2 \tag{4.1}$$

where F_0 represents axial compressive load and L(w) – a nonlinear part of elastic forces due to large deflections (geometric nonlinearity) and assume a single-mode solution

$$w(y,\tau) = \psi_1(y)\,\check{x}(\tau) \; ; \quad \psi_1(y_0) = 1 \; ; \tag{4.2}$$

where $\psi_1(y)$ is a fundamental eigenfunction (normal mode shape) of the linear system governed by equations

$$m\frac{\partial^2 w}{\partial \tau^2} + EJ\frac{\partial^4 w}{\partial y^4} = 0 \, ,$$
$$B_{i1}(w)_{y=0} = 0 \; ; \quad B_{i2}(w)_{y=\ell} = 0 \; ; \tag{4.3}$$

On inserting eqs. (4.2) into eqs. (4.1) and applying the Galerkin procedure one arrives at a single second order differential equation for the normal coordinate $\check{x}(t)$. In the post-buckling conditions $F_0 > F_{0cr}$ the equation transformed into a nondimensional form is written as [1,3,13] :

$$\ddot{x} + \gamma\dot{x} - \frac{1}{2}(1 - x^2)x = f\cos\omega t \, ,$$
$$x = \frac{\check{x}}{\ell} \; ; \tag{4.4}$$

where t and is a nondimensional time, and the nonlinear term $1/2x^3$ is, in general case, the first term in Taylor expansion of the function representing nonlinear part of elastic force.

From the theory of nonlinear vibrations it is known that the system (4.4) at $f = 0$ has three equilibrium positions:

$x_{(1)} = 0$ – unstable position (saddle point in $x - \dot{x}$ plane);

$\left.\begin{array}{l} x_{(2)} = +1 \\ x_{(3)} = -1 \end{array}\right\}$ – stable positions (stable foci in $x - \dot{x}$ plane);

and that the complete system may exhibit two types of periodic steady-state oscillations: around the stable equilibrium positions, called later "Small Orbit" and oscillations around all three equilibrium positions, called "Large Orbit" (see Fig. 4.1). It depends on initial conditions which type of motion is really generated in the system.

For better insight into the Small Orbit motion one may shift the origin of coordinates on introducing new variable

$$z = x + x_{(2),(3)} \quad , \tag{4.5}$$

and transform eqs. (4.4) into the form

$$\ddot{z} + \gamma \dot{z} + z \mp \frac{3}{2} z^2 + \frac{1}{2} z^3 = f \cos \omega t \; ; \tag{4.6}$$

Thus it is seen that the natural linear frequency of the Small Orbit is equal to one: $\omega_0 = 1$.

Indeed the system governed by eqs. (4.4) behaves that way i.e. exhibits periodic oscillations either of Small Orbit or Large Orbit type provided the amplitude of forcing function f at given γ does not exceed a certain critical value. One may expect that on increasing the parameter f/γ Small Orbit motion would jump into Large Orbit oscillations – the jump phenomena belong to the most characteristic features of nonlinear vibrating systems. However, jump phenomena of this type do not always occur. Instead when f/γ exceeds a critical value Small Orbit turns into an irregular motion consisting essentially of random-like jumps from oscillations around $x_{(2)} = +1$ to oscillations around $x_{(3)} = -1$. The motion does not decay and in a long time interval shows properties of a "steady-state" behaviour. This is shown in Fig. 4.2. where chaotic motion is illustrated by means of three descriptors: time history, phase portrait and Poincare map.

Experimental results presented by Moon [13] allow one to make the observation that critical values of the parameter f/γ for the change of Small Orbit into the chaotic motion take minimum values in the region of frequency close to the principal resonance i.e. at $\omega \approx 1.0$. In this section attention is focused on this zone of the excitation frequency.

4.2. The First Approximate Solution and Peculiar Features of Resonance Curves

In the neighbourhood of the principal resonance of Small Orbit and at sufficiently low values of the parameter f/γ it is legitimate to seek the first approximate periodic solution as (see e.g. [8]):

$$x(t) = A_0 + A_1 \cos(\omega t + \vartheta_1) = x(t + T) \; ,$$
$$T = 2\pi/\omega \quad ; \tag{4.7}$$

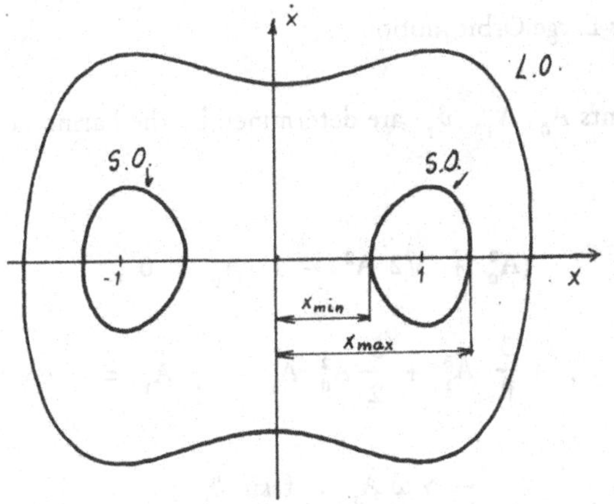

Fig. 4.1. Phase portrait of Small and Large Orbit motion.

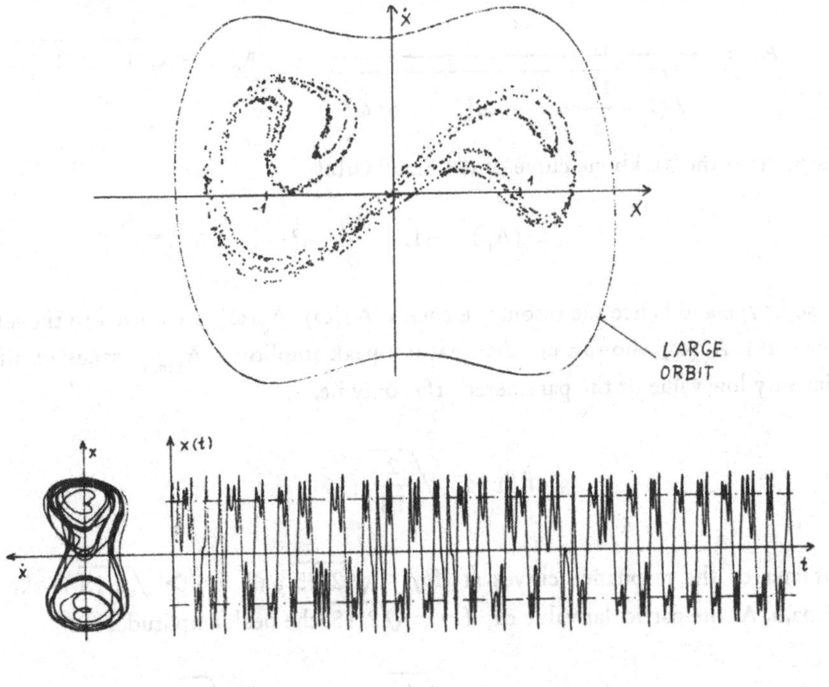

Fig. 4.2. Time history, phase portrait and Poincare map (strange attractor) of chaotic motion: $f = 0.14$; $\gamma = 0.10$; $\omega = 0.8$.

where $A_0 \neq 0$ for Small Orbit solution and

$A_0 = 0$ for Large Orbit motion.

The unknown coefficients A_0, A_1, ϑ_1 are determined by the harmonic balance method i.e. they satisfy equations

$$(A_0^2 + 3/2\, A_1^2 - 1)\, A_0 = 0\,,$$

$$-\omega^2 A_1 + \frac{3}{8} A_1^3 + \frac{3}{2} A_0^2 A_1 - \frac{1}{2} A_1 = f \cos\vartheta_1\,, \tag{4.8}$$

$$-\gamma\omega A_1 = f \sin\vartheta_1\,;$$

Equations (4.8) are derived on inserting eqs. (4.7) into eqs. (4.4) and equating a constant term and coefficients of $\cos(\omega t + \vartheta_1)$ and $\sin(\omega t + \vartheta_1)$ separately to zero. Solving of eqs. (4.8) gives the sought amplitudes A_1, A_0 as functions of frequency at $A_0 \neq 0$:

$$A_1 = \frac{f}{\sqrt{(1 - \frac{15}{8} A_1^2 - \omega^2)^2 + \gamma^2\omega^2}}\;; \qquad A_0 = \mp\sqrt{1 - 3/2\, A_1^2} \tag{4.9}$$

It comes out that the backbone curve of the Small Orbit:

$$\bar{\omega}(A_1) = 1 - \frac{15}{8} A_1^2\,, \tag{4.10}$$

is of the soft type and hence the resonance curves $A_1(\omega)$, $A_0(\omega)$ are bowed to the left. The resonance curve $A_1(\omega)$ shows a peculiar feature: peak amplitude $A_{1\,max}$ takes a finite value for sufficiently low value of the parameter f/γ only i.e. at

$$f/h < \sqrt{\frac{2}{15}}\;; \tag{4.11}$$

The character of the resonance curves at $f/\gamma < \sqrt{2/15}$ and $f/\gamma > \sqrt{2/15}$ is sketched in Fig. 4.3a,b. At the particular value of $f/\gamma = \sqrt{2/15}$ the peak amplitudes are:

$$A_1 \equiv A_1(\omega_B) = \sqrt{\frac{4}{15}}\;; \qquad \omega_B = \sqrt{\frac{1}{2}}\;; \tag{4.12}$$

While the resonance curve in Fig. 4.3a takes the classical shape of that in the Duffing equation with soft nonlinearity, the one in Fig. 4.3b looks like a resonance curve in an undamped system.

Since the system considered is damped ($\gamma > 0$) the only reasonable explanation of the peculiar behaviour of $A_1(\omega)$ is that the first approximate harmonic solution (4.7 is not adequate one for so high value of the parameter f/γ.

In spite of this observation, we study the harmonic solution at $f/\gamma > \sqrt{2/15}$ because surprising and promising results are obtained where instead of the amplitudes A_1 and A_0 a maximum and minimum of displacement is considered:

$$A_0 + A_1 \equiv A_{max}(\omega) ; \qquad A_0 - A_1 = A_{min} \equiv A_{min}(\omega) ;$$

Indeed resonance curves $A_{max}(\omega)$ and $A_{min}(\omega)$ appear to show a behaviour close to x_{max} and x_{min} obtained by computer simulation even for large values of the parameter f/γ and become a key to obtaining an approximate criterion for chaotic motion to appear.

At $f/\gamma > \sqrt{2/15}$ a characteristic feature of the resonance curve $A_0 + A_1 \equiv A_{max}(\omega)$ is an appearance of a point with horizontal tangent. Coordinates of the point, denoted by D in Fig. 4.3c, are derived as follows: since $dA_1/d\omega \neq 0$ and has a finite value on the resonant branch of $A_1(\omega)$, the condition

$$\frac{d}{d\omega}(A_0 + A_1) = \frac{d}{dA_1}(A_0 + A_1)\frac{dA_1}{d\omega} = 0 , \tag{4.13a}$$

is equivalent to the condition

$$\frac{d(A_0 + A_1)}{dA_1} = 0 ; \tag{4.13b}$$

The latter is easily derived by virtue of eqs. (4.8) with the results

$$A_1(\omega_D) = \sqrt{\frac{4}{15}} ; \qquad A_0(\omega_D) = \sqrt{\frac{3}{5}} ;$$

$$A_{max} = A_1(\omega_D) + A_0(\omega_D) \approx 1.29 ; \tag{4.13c}$$

$$\omega_D^2 = \frac{1}{2}[1 - \gamma^2 + \sqrt{\gamma^4 - 2\gamma^2 + 15 f^2}] ;$$

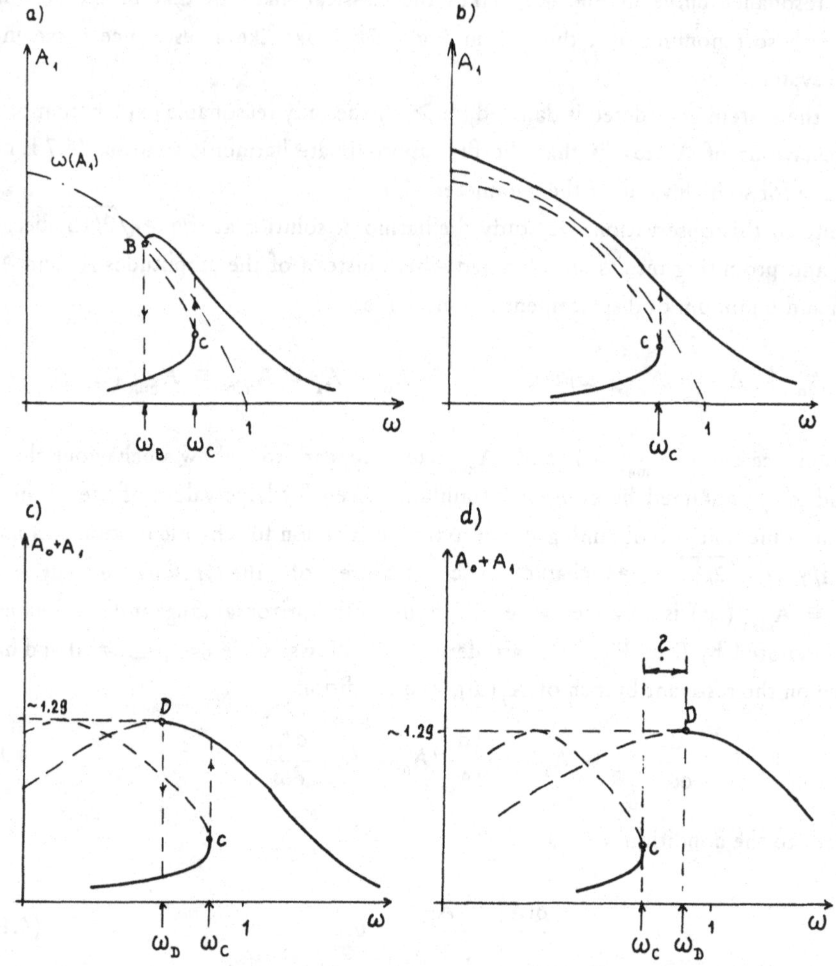

Fig. 4.3. Resonance curves of the approximate harmonic solution.

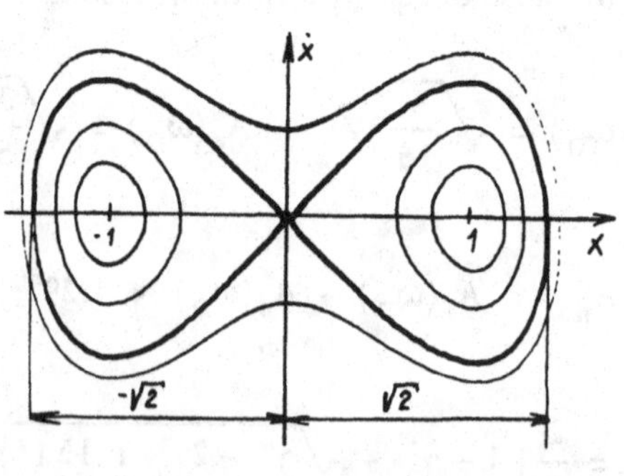

Fig. 4.4. Phase portrait of hamiltonian system.

It is worth noticing that the value $A_{max}(\omega_D)$ thus calculated is constant, independent of the parameters f and γ. The observarion brings immediately an idea to compare the result with the displacement of the Small Orbit in the Hamiltonian system governed by the equation

$$\ddot{x} - \frac{1}{2}(1 - x^2)x = 0 \; ; \tag{4.14a}$$

One can easily find out that at the separatrix, which is a boundary line between Small and Large Orbit motion, the maximum deflection is (see Fig. 4.4):

$$x^{(h)}_{max} = \sqrt{2} \equiv 1.41 > A_{max}(\omega_D) \; ; \tag{4.14b}$$

This is what one expects intuitively: the forced Small Orbit of the complete system (4.4) should remain inside the separatrix loop of the Hamiltonian system (see also [13]).

The other characteristic point in Fig. 4.3c,d – the point of the vertical tangent – C is the classical stability limit of the harmonic solution. The point lies on the non resonant branch of the resonance curve and hence its dependence on the damping coefficient at the low values of γ considered is negligible. At $\gamma = 0$ the frequency is given by

$$\omega_c^2 = 1 - \frac{3}{2}\sqrt[3]{\frac{15}{4}\, f^2} \; ; \tag{4.15}$$

Although one cannot prove that the point with horizontal tangent (point D) is also a stability limit, physical intuition says that the resonant branch of the maximum displacement $A_{max}(\omega)$, where $dA_{max}/d\omega > 0$ should not be realized in real systems. On using this argument the branch of $A_{max}(\omega)$ on the left from the point D has been denoted as "unstable" branch in Fig. 4.3c,d. Therefore as long as

$$\omega_D > \omega_c ,$$

as it is sketched in Fig. 4.3c, the stable harmonic solution of the Small Orbit exists in the whole range of frequency considered and the classical jump phenomena from non resonant to resonant branch of the resonance curve at $\omega = \omega_c$, on increasing ω, and from resonant to non resonant solution at point $\omega = \omega_D$ on decreasing frequency, can be expected.

When the forcing parameter is further increased, however, the frequency ω_D grows and ultimately exceeds the frequency ω_D (see Fig. 4.3d). In this case a stable harmonic solution for the Small Orbit does not exist in the region $\omega_c - \omega_D$ and an appearance of "strange phenomena" can be expected on reaching frequency ω_c or ω_D. This observation

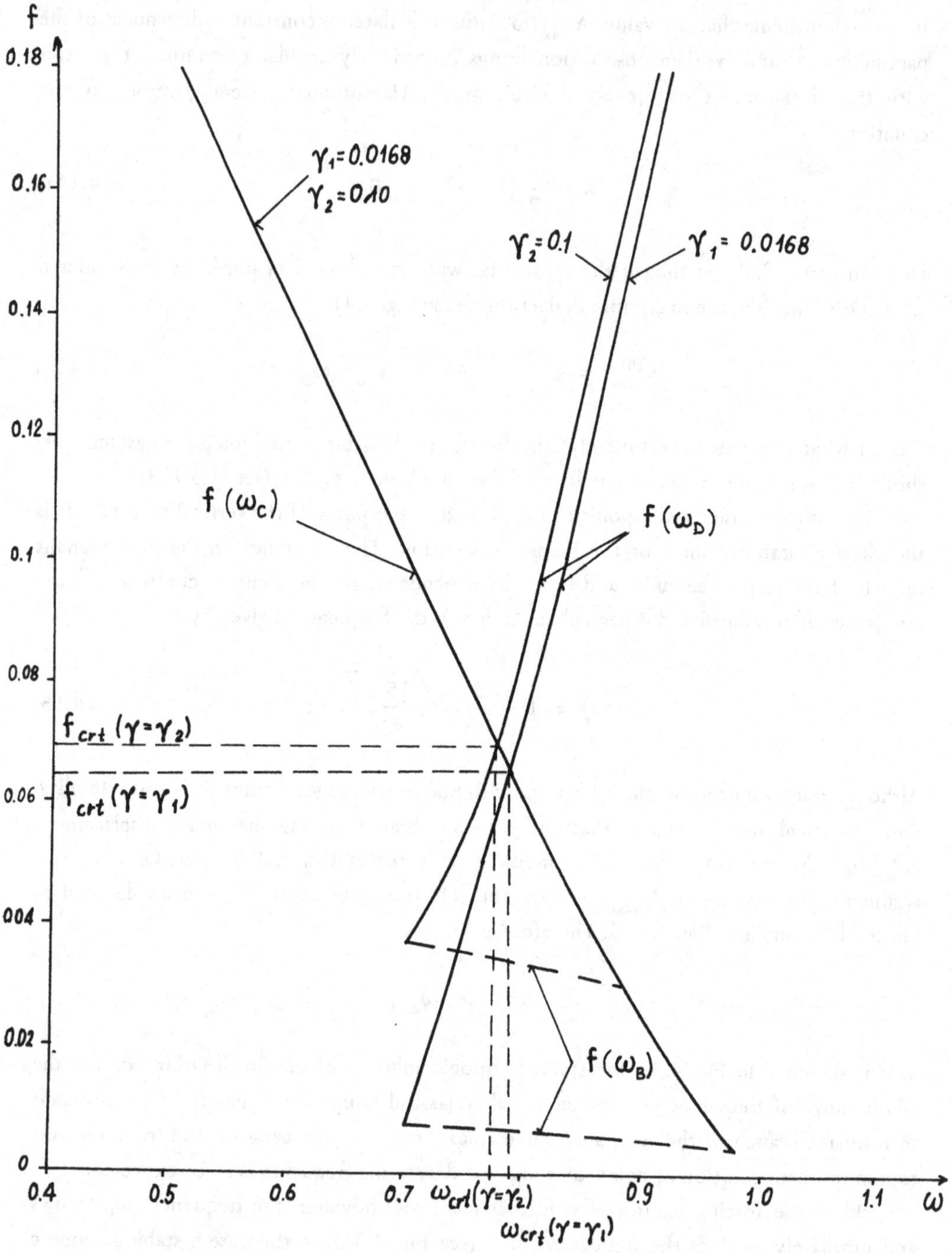

Fig. 4.5. Loci of characteristic points B,C,D (denoted in Fig. 4.3.) on $f - \omega$ plane at γ_1 =0.0168 and γ_2 = 0.10

makes an essential point in the attempt of proposing an approximate criterion for chaotic motion based on the first approximate solution.

Frequencies of the two characteristic points ω_c and ω_D evaluated by eqs. (4.13c) and (4.15) are displayed on $f-\omega$ plane at two values of the damping coefficient (see Fig. 4.5). Indeed the two curves $f \equiv f(\omega_c)$ and $f \equiv f(\omega_D)$ cross each other at a point whose coordinate have been denoted by f_{crt} and ω_{crt}. By virtue of eqs. (4.13c) and (4.15) the critical point satisfies relations

$$\omega_c^2 = \omega_D^2 ,$$

$$1 - \frac{3}{2}\sqrt[3]{\frac{15}{4} f_{crt}^2} = \frac{1}{2}[1 - \gamma^2 + \sqrt{\gamma^4 - 2\gamma^2 + 15 f_{crt}^2}\,] ; \tag{4.16a}$$

or, in a reduced form as

$$\omega_{crt}^2 = 1 - 2.33\, \psi_{crt}^{2/3} ;$$

$$4\gamma^2(1 - 2.33\, f_{crt}^{2/3}) = 15 f_{crt}^2 - (1 - 4.66\, f_{crt}^{3/2})^2 \tag{4.16b}$$

It is clearly seen that the region where "strange phenomena" can be suspected is defined by equations

$$f > f_{crt}$$

$$\omega_c < \omega < \omega_D \tag{4.16c}$$

Loci of the other characteristic point on the resonance curve which appears only at $f/\gamma < \sqrt{2/15}$ – point B in Fig. 4.3a – are also drawn in Fig. 4.5.

4.3. Computer Simulation Analysis

To verify the theoretical predictions and to investigate true behaviour of the system equation of motion (4.4) was simulated on both analog and digital computers and response of the system was observed, transformed and recorded. Simulation results are displayed in Fig. 4.2, 4.6, 4.12.

First in Fig. 4.6, periodic motion is characterized by resonance curves of maximal and minimal displacements of Small Orbit – $X_{max}(\omega)$ and $X_{min}(\omega)$ (see Fig. 4.1) drawn together with the theoretical values $A_{max}(\omega)$ and $A_{min}(\omega)$, respectively. Minimal displacement of Large Orbit motion, generated in the system on applying appropriate,

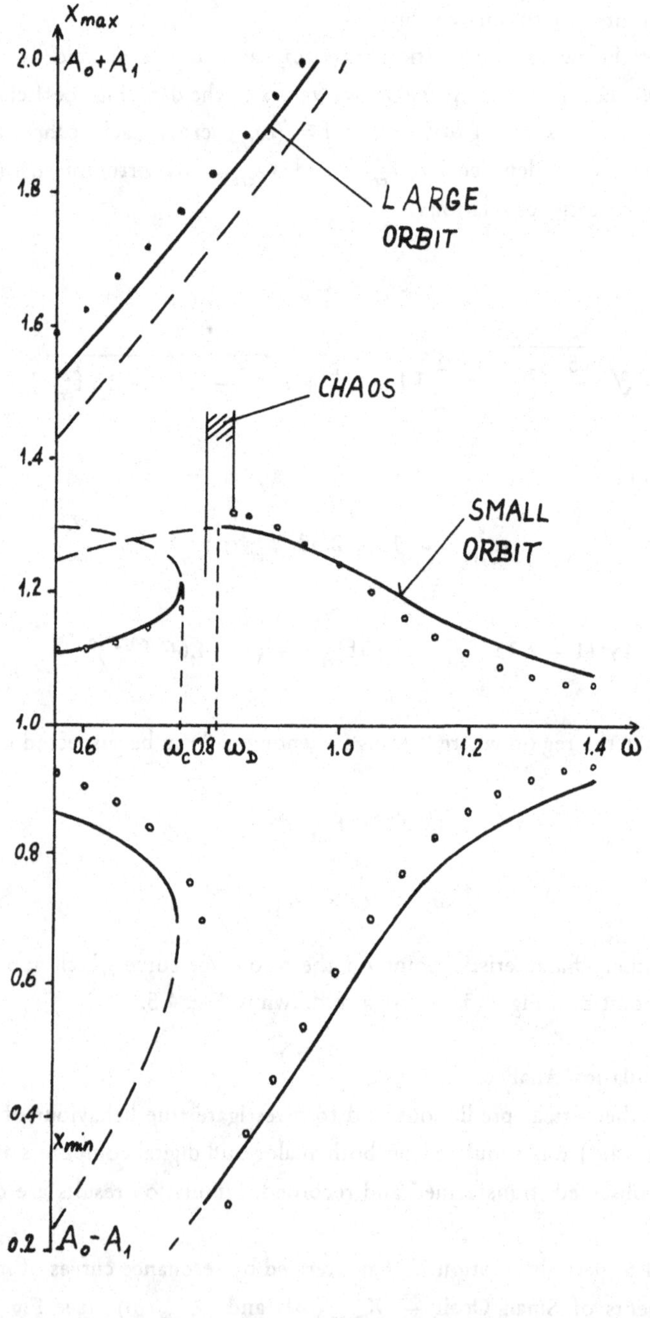

Fig. 4.6. Resonance curves of maximal and minimal displacement in Small and Large Orbit motion at f = 0.08; γ = 0.0168; ——— – – – theoretical, ○ ○ ● ● simulation results.

sufficiently large initial conditions, is also shown together with amplitude of harmonic solution $A_1(\omega)$ at $A_0 = 0$.

It is readily seen that the theoretical and simulation results show, in general, good agreement, in spite of rather high value of the forcing parameter satisfying the condition $\omega_D > \omega_C$. Computer simulation confirms the theoretical prediction that "strange phenomena" may appear in this case. Indeed chaotic behaviour was observed in the region of frequency denoted as CHAOS in Fig. 4.6. The region is very close to the theoretical zone where the stable Small Orbit harmonic solution does not exist i.e. the zone defined by eqs. (4.16c).

To illustrate various types of response close to, and inside the chaotic zone, time histories, phase portraits and Poincare maps were recorded at several values of frequency and displayed in Fig. 4.7a-e. First Small Orbit motion at frequency near the top of $X_{max}(\omega)$ curve i.e. at $\omega = 0.85$ and $\omega = 0.835$ slightly higher than ν_D is shown in Fig. 4.7ab. This is the region of frequency where the simulation value X_{max} is clearly higher than the theoretical A_{max} amplitude. The time history, phase portrait and harmonic analysis of x(t) in Fig. 4.7a give reason for the discrepancy. The response x(t) is still T-periodic here, but the second harmonic component appears to take relatively large value and one may conclude that a legitimate approximate solution at $\omega \approx 0.85$ is

$$x(t) = A_0 + A_1 \cos(\omega t + \vartheta_1) + A_2 \cos(2\omega t + \vartheta_2) ; \tag{4.17}$$

The response at slightly lower frequency $\omega = 0.835$ in Fig. 4.7b s no longer T-periodic, but appears to be 2T-periodic (two points on Poincare map). Then the 2T-periodic motion turns into chaotic motion at $\omega = 0.78$ (see Fig. 4,7c). Due to low damping the strange attractor does not show a clear-cut structure, but time history and phase portrait are the same character as those in Fig. 4.2 obtained at higher value of the damping coefficient $\gamma = 0.10$. On further decrease of frequency chaotic motion turns rapidly into the T-periodic Small Orbit motion (see Fig. 4.7d). It is worth to point out that no hysteresis in the system behaviour was observed – responses were the same whether frequency was decreasing or increasing. Large Orbit motion shown in Fig. 4.7e was generated in the system on applying appropriate, sufficiently large initial conditions only.

This type of the system behaviour – transition from resonant to non resonant Small Orbit, or vice versa, through chaotic motion zone was observed for a certain region of the forcing parameter i.e. at (see Fig. 4.11 and 4.12):

$$f_{cr} < f < f_{cr}^{(h)}$$

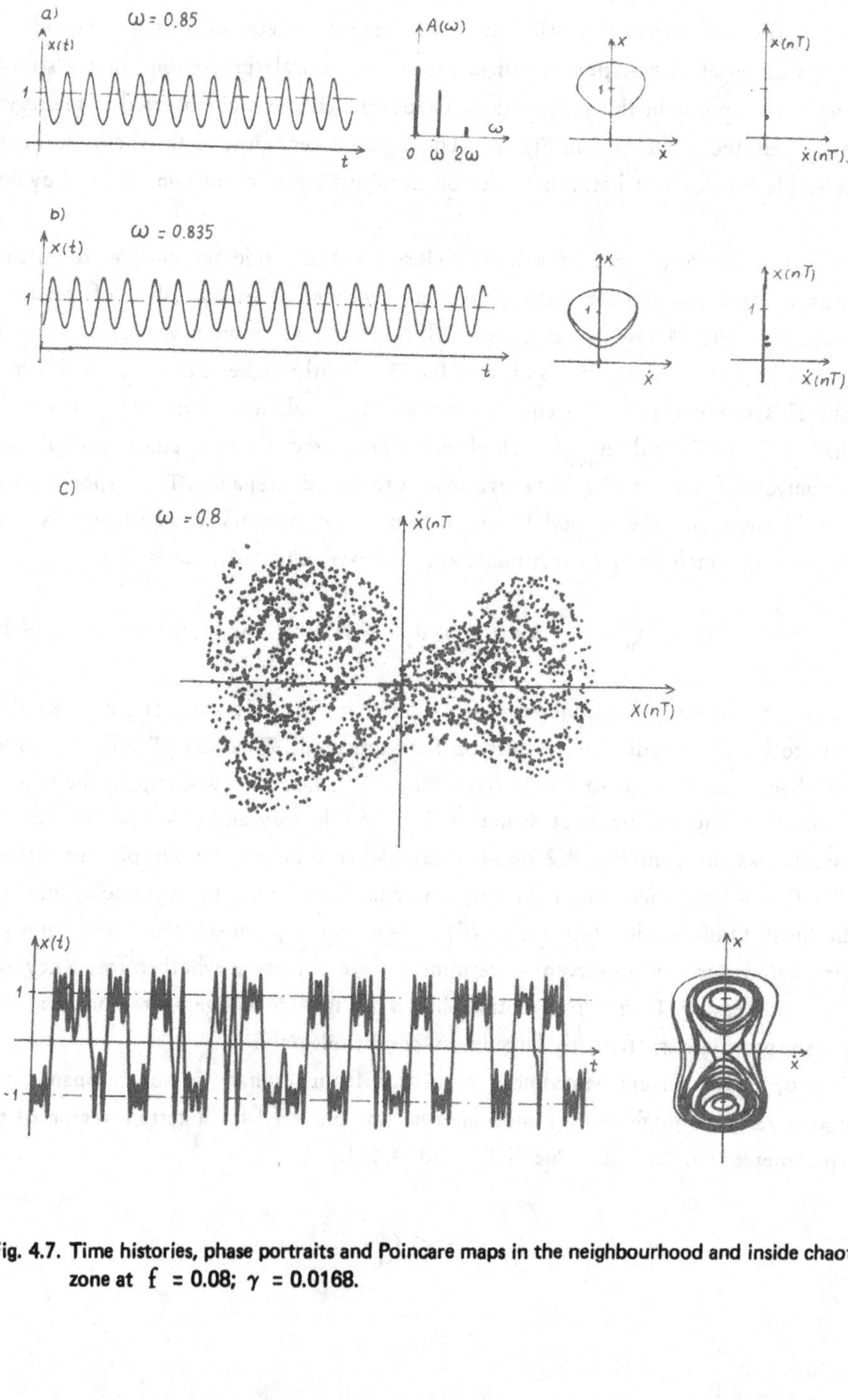

Fig. 4.7. Time histories, phase portraits and Poincare maps in the neighbourhood and inside chaotic zone at f = 0.08; γ = 0.0168.

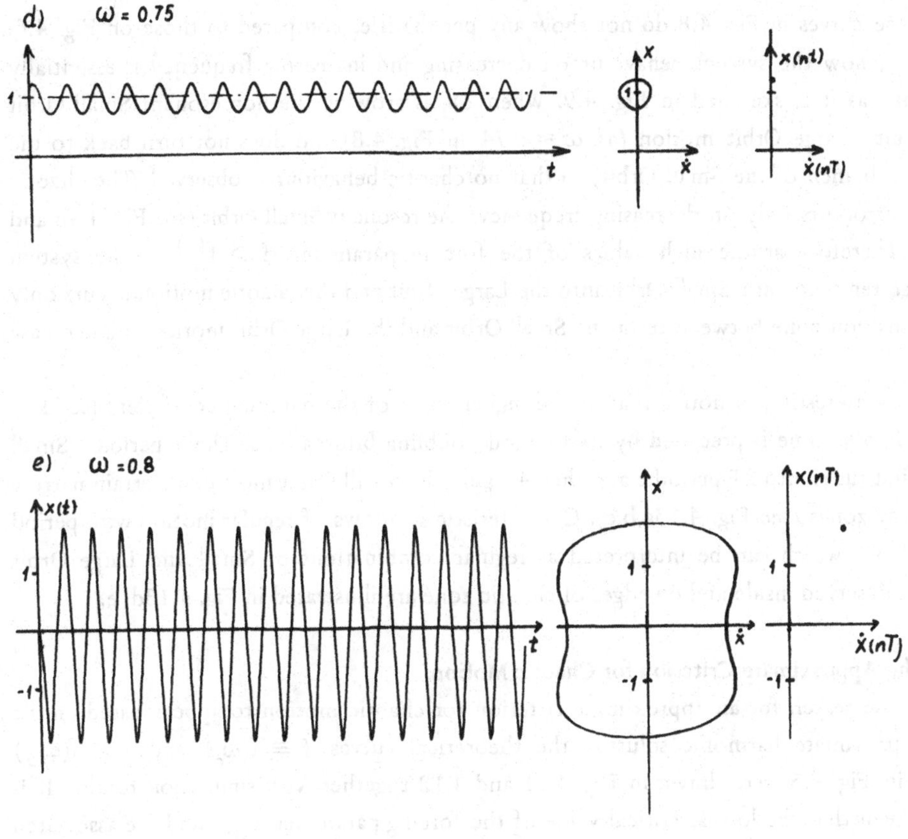

Fig. 4.7. continued

At $f > f_{cr}^{(h)}$ a distinctly new phenomenon was noticed – a hysteresis type behaviour accompanied with a jump from Small Orbit to Large Orbit motion. This is illustrated in Figs. 4.8 – 4.10. First we notice that resonance curves in Fig. 4.8 do not show any peculiarities compared to those on Fig. 4.6. However, now the system behaviour on decreasing and increasing frequency is essentially different, as it is sketched in Fig. 4.9. When ω is growing the nonresonant Small Orbit jumps into Large Orbit motion (at $\omega \approx 0.74$ in Fig. 4.8) and does not turn back to the resonant branch of the Small Orbit, so that no chaotic behaviour is observed. The chaotic behaviour occurs only on decreasing frequency: the resonant Small Orbit (see Fig. 4.9b and 4.10). Therefore at the high values of the forcing parameter $f > f_{cr}^{(h)}$ the system response tends to turn Small Orbit into the Large Orbit and the chaotic motion occurs only as a transition zone between resonant Small Orbit and the Large Orbit motion at a decrease of ω.

It is interesting to notice that at the higher value of the damping coefficient (see Fig. 4.12) chaotic zone is preceded by two period doubling bifurcations. The T-periodic Small Orbit first turns into 2T-periodic and then 4T-periodic Small Orbit motion in certain narrow frequency zones (see Fig. 4.13a,b,c). Characteristic windows of regular motion with period 3T and 5T, which can be interpreted as regular combinations of Small and Large Orbit motion, observed inside and on edges of chaotic zone are illustrated in Fig. 4.13d, e.

4.4. The Approximate Criterion for Chaotic Motion

In the search for an approximate criterion for chaotic motion to appear based on the first approximate harmonic solution the theoretical curves $f \equiv f(\omega_c)$ and $f \equiv f(\omega_D)$ shown in Fig. 4.5 were drawn in Fig. 4.11 and 4.12 together with simulation results. It is clearly seen that the lowest critical value of the forcing parameter f_{crt} and the associated ω_{crt} are very close to those obtained by computer simulation. This is an essential observation: in spite of the fact that the true response close to chaotic zone differs considerably from harmonic motion the theoretical lowest value of the forcing parameter f_{crt}, when chaotic behaviour is predicted, shows surprisingly good coincidence with the true value f_{cr} (see Fig. 4.14).

At higher values of the parameter f, the theoretical range of frequencies where Small Orbit motion was predicted to disappear is shifted to the left compared with computer simulation results. Still a coincidence of the results is so good that the region on $f - \omega$ plane determined by eqs, (4.16c) is proposed as the approximate criterion for chaotic motion at the forcing parameter slightly exceeding f_{crt} .

Conclusions

The study of a mathematical model of a buckled beam by the use of the approximate

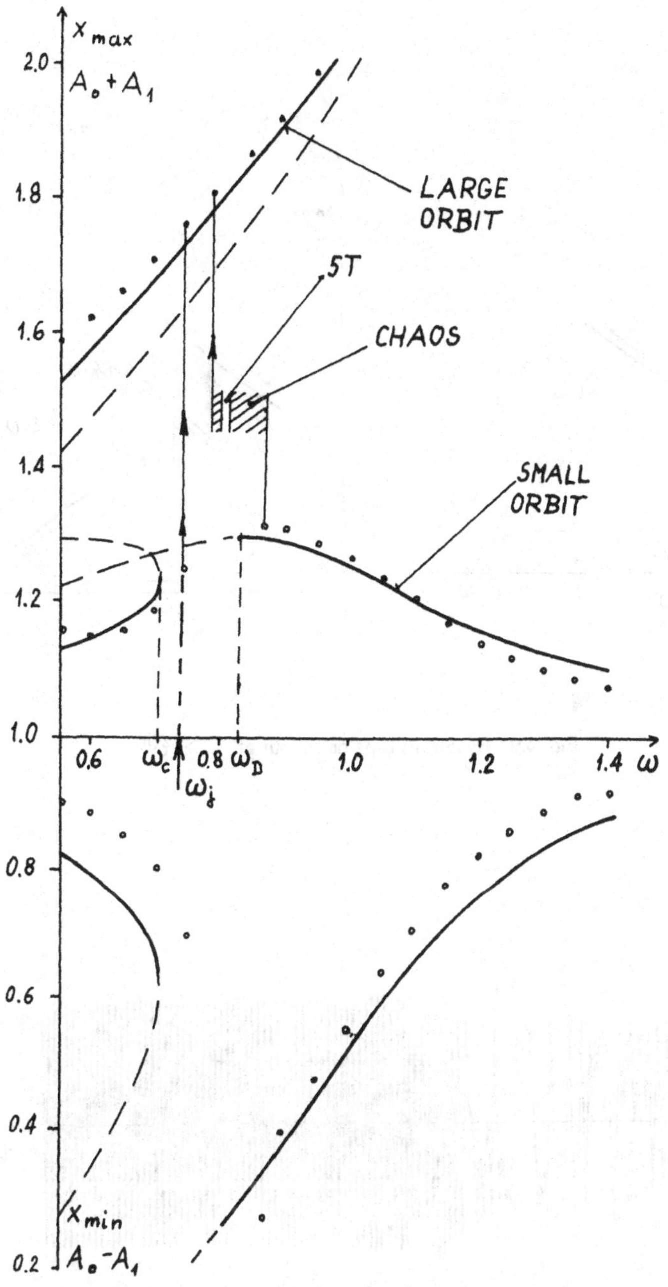

Fig. 4.8. Resonance curves of maximal and minimal displacement in Small and Large Orbit motion at f = 0.10; γ = 0.0168; ——— - - - theoretical, ○○○● ● simulation results.

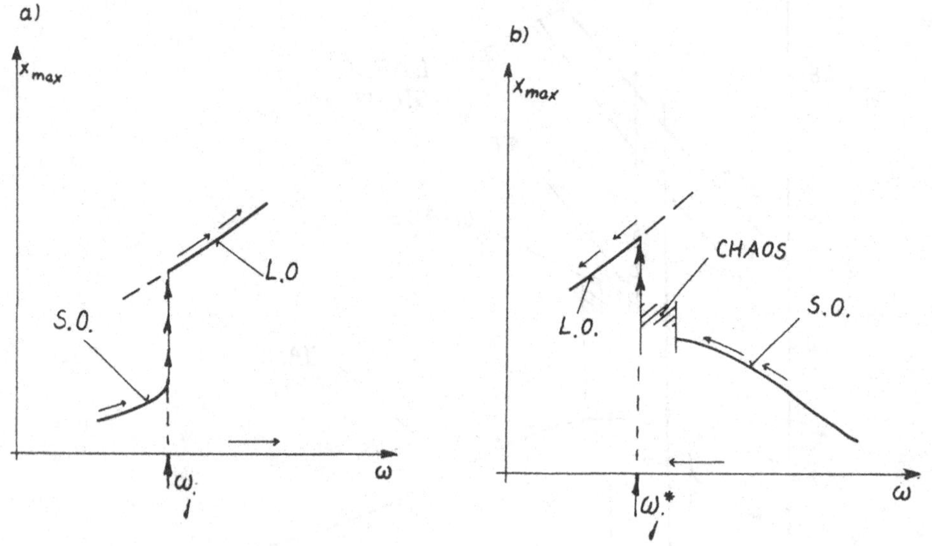

Fig. 4.9. Hysteresis type behaviour at $f > f_{cr}^{(h)}$

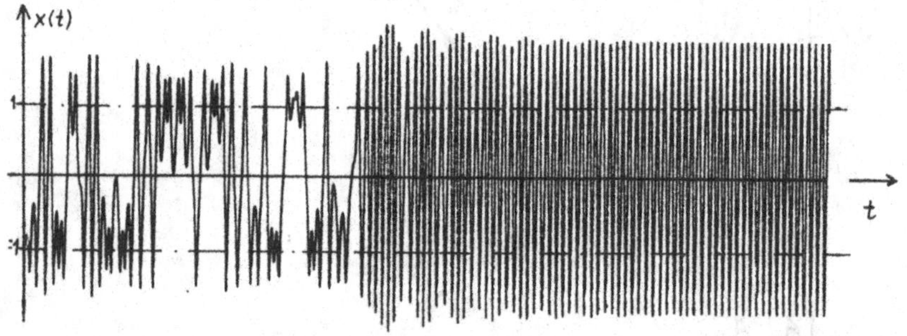

Fig. 4.10. Time history of transient between chaotic and Large orbit motion at $\omega = \omega_j^*$.

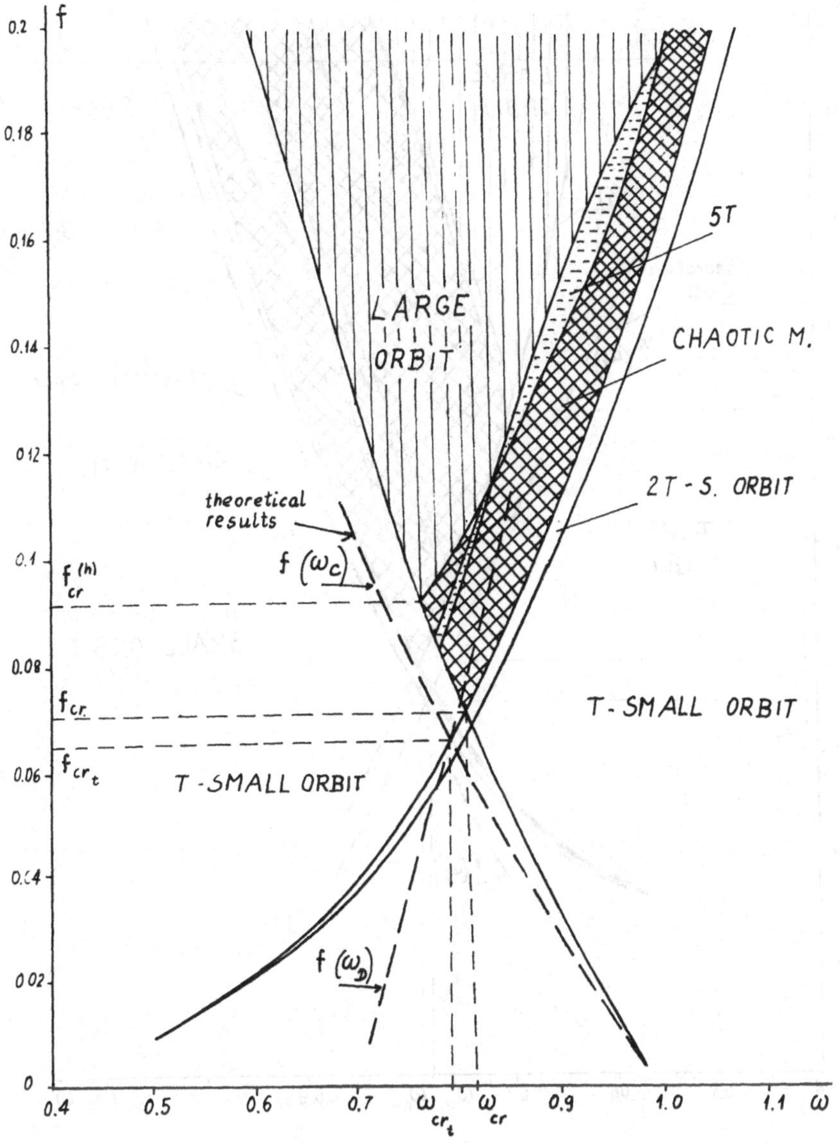

Fig. 4.11. Regions of various periodic motion and chaotic behaviour in $f - \omega$ plane at $\gamma = 0.0168$ – simulation results. – – – theoretical curves $f = f(\omega_c)$ and $f = f(\omega_D)$.

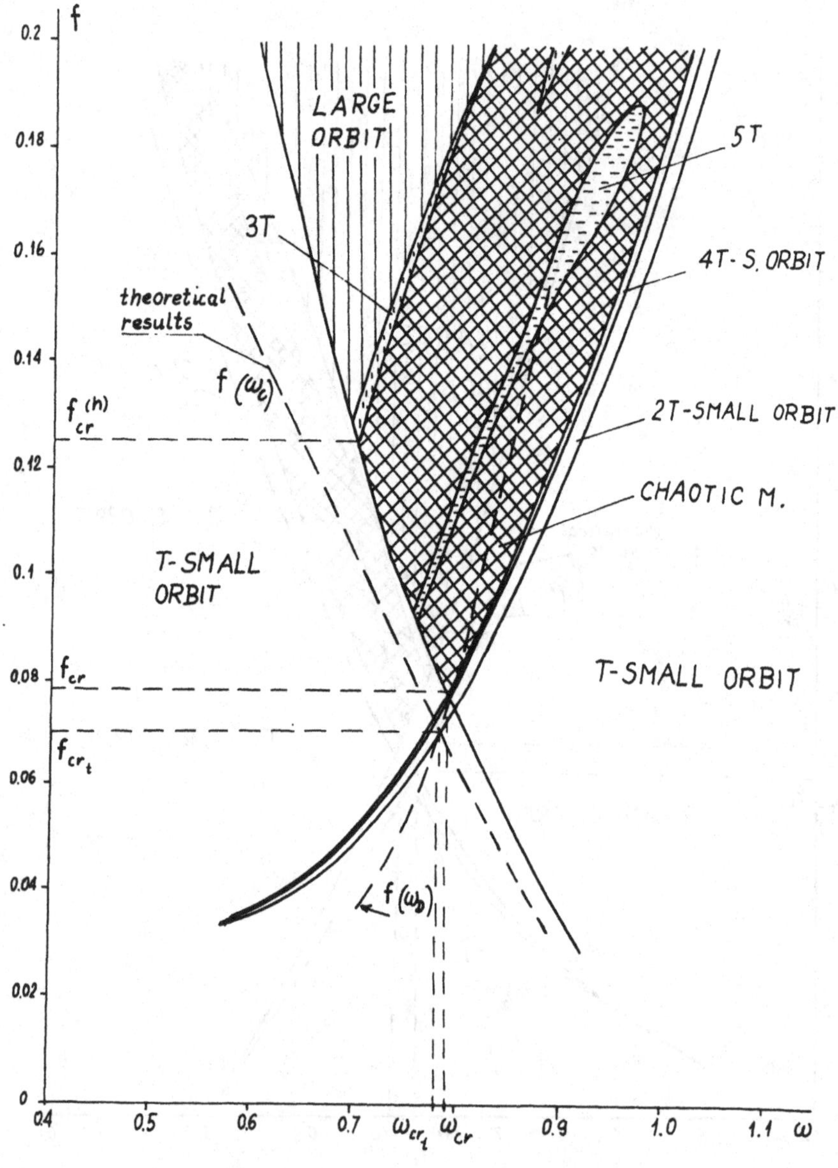

Fig. 4.12. Regions of various periodic motion and chaotic behaviour at $\gamma = 0.10$ – computer simulation results; – – – theoretical $f \equiv f(\omega_c)$ and $f \equiv f(\omega_D)$ curves.

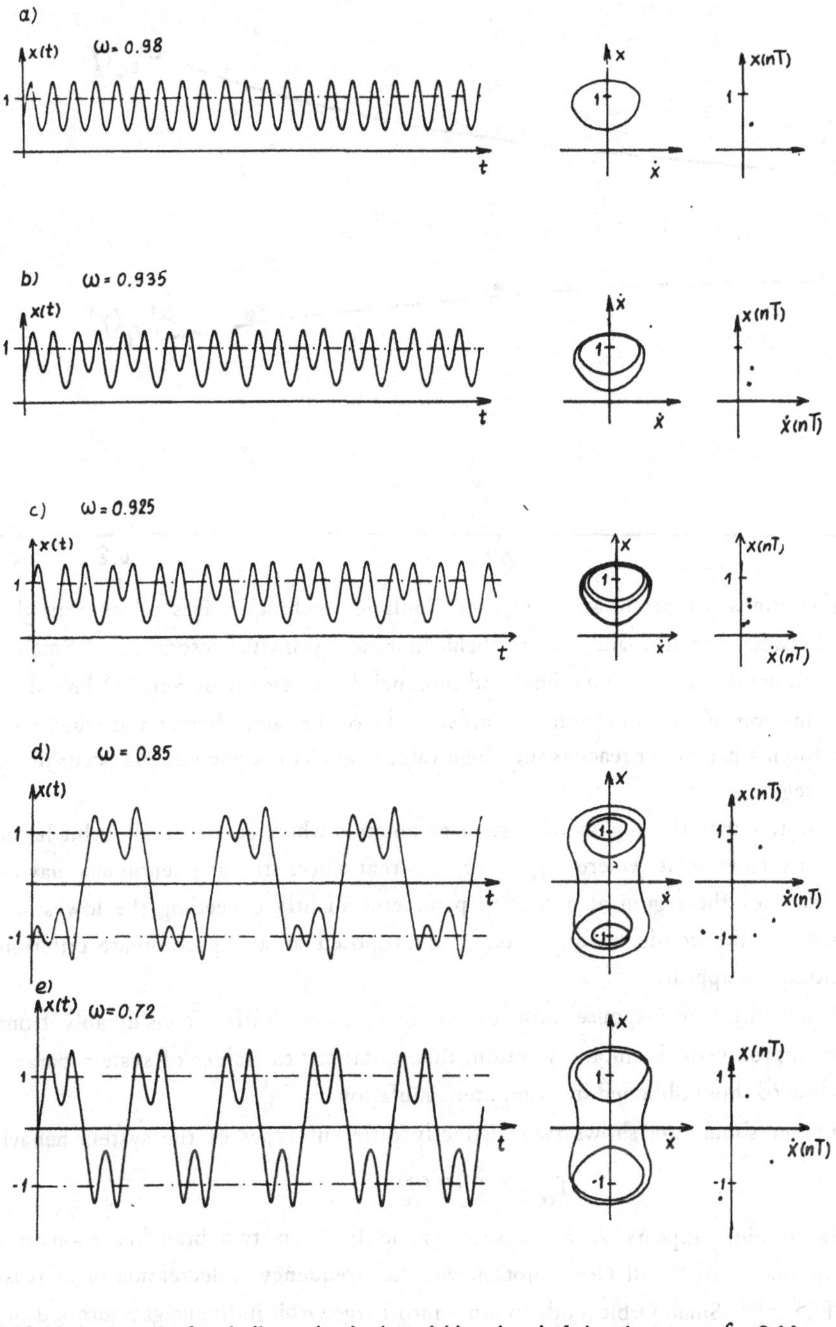

Fig. 4.13. Various types of periodic motion in the neighbourhood of chaotic zone at $f = 0.14$; $\gamma = 0.10$.

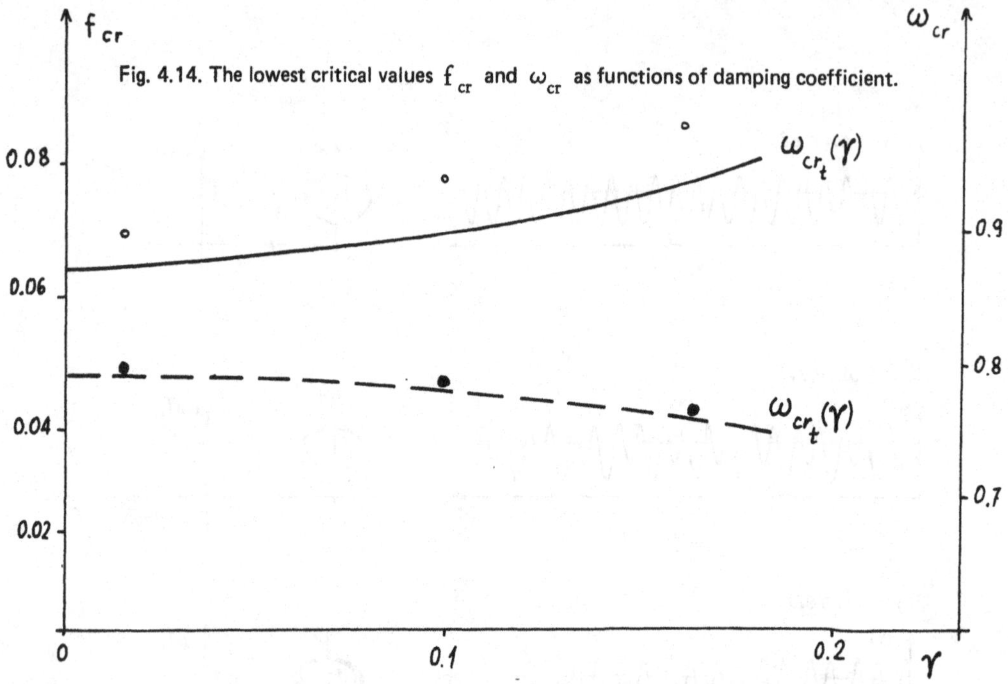

Fig. 4.14. The lowest critical values f_{cr} and ω_{cr} as functions of damping coefficient.

theory of nonlinear vibrations and computer simulation technique gives a deeper insight into relations between periodic and chaotic behaviour near principal resonance of Small Orbit motion. Resonance curves of maximal and minimal displacement of Small Orbit calculated on an assumption of harmonic solution prove to be of the same character as true ones even when the forcing parameter reaches such high value that chaotic phenomena occur in certain frequency region.

It is stated that the region of system parameters where stable Small Orbit harmonic solution cease to exist i.e. where $\omega_D > \omega_c$, is that where strange phenomena may occur. The boundary of the region at a forcing parameter slightly exceeding the lowest critical value satisfying the condition $\omega_D = \omega_c$ is proposed as an approximate criterion for chaotic motion to appear.

Although the true response close to the chaotic zone differs considerably from the T-periodic approximate harmonic solution, theoretifal critical values of system parameters are very close to those obtained by computer simulation.

Computer simulation shows two distinctly different types of the system behaviour: first at

$$f_{cr} < f < f_{cr}^{(h)}$$

the chaotic motion appears as a transition zone between two branches-resonant and nonresonant one – of Small Orbit motion whether frequency is decreasing or increasing. Then at $f > f_{cr}^{(h)}$ Small Orbit tends to jump into Large Orbit in the neighbourhood of the principal resonance considered. Chaotic motion occurs as a transition zone between resonant Small Orbit and Large Orbit on decreasing frequency only.

REFERENCES

[1] W. W. Bolotin, Dynamic Stability of Elastic Systems, San Francisco, Holden Day, 1964.

[2] P. Holmes, Strange phenomenon in dynamical systems and their physical implications, Appl. Math. Modelling, 1, 1977, 362-366.

[3] P. Holmes, A nonlinear oscillator with a strange attractor, Phil. Trans. of the Royal Society of London, 292, 1977, 419-448.

[4] P. Holmes, Averaging and chaotic motion in forced oscillations, SIAM, J. Appl. Math., 38, 1980, 65-80.

[5] C. Holmes and P. Holmes, Second order averaging and bifurcation to subharmonics in Duffing's equation, 1981, J. Sound and Vibration, 78, 2, 161-174.

[6] P. Holmes and F. C. Moon, Strange attractors and chaos in nonlinear mechanics, J. Appl. Mech., 50, 1983, 1021-1032.

[7] J. Guckenheimer and P. Holmes, Nonlinear oscillations, Dynamical Systems and bifurcations of vector fields, Springer Verglag, 1983.

[8] Ch. Hayashi, Nonlinear oscillations in physical systems, McGraw Hill Book Co., New York, 1964.

[9] G. Iooss and D.D. Jospeh, Elementary stability and bifurcation theory, Springer Verlag, New York, 1981.

[10] R. W. Leven, B. Pompe, C. Wilke and B. P. Koch, Experiments on periodic and chaotic motions of a parametrically forced pendulum, Physica 16d, 1985, 371-384.

[11] S. Maezawa, Steady forced vibrations of unsymmetrical piecewise linear systems, Bull. JSME 4, 1980, 200-229.

[12] S. Mazeawa, H. Kumano and Y. Minakuchi, Forced vibrations in an unsymmetric linear system excited by general periodic forcing functions, Bull. JSME 23, 1980, 68-75.

[13] F. C. Moon, Experiments on chaotic motions of a forced nonlinear oscillator: strange attactors, J. Appl. Mech. 47, 3,1980, 638-643.

[14] K. Popp, Chaotische Bewegungen beim Duffing Schwinger, Festschrift zum 70 Geburtstag von Herrn prof. K. Magnus, München 1982.

[15] J. Rudowski and F. C. Moon, Chaos in nonlinear mechanics, Reports of the Institute of Fundamental Technological Research, 28, 1985.

[16] J. Rudowski and W. Szemplinska-Stupnicka, On an approximative criterion for chaotic motion in a model of a buckled beam, Ingenieur Arch., / to appear /.

[17] H. G. Schuster, Deterministic chaos an introducing, Physic-Verlag 1984.

[18] R. Seydel, The strange attractors of a Duffing equation dependence on the exciting frequency, Report of TU-München, Inst. f. Mathematik, TUM-M8019, 1980.

[19] S. W. Shaw and P. J. Holmes, A periodically forced piecewise linear oscillator, J. Sound and Vibration, 90/1/, 1983, 129-155.

[20] W. Szemplinska-Stupnicka, On higher harmonic oscillations in heteronomous nonlinear systems with one degree of freedom, Int. J. Non-linear Mechanics 3, 1968, 17-30.

[21] W. Szemplinska-Stupnicka and J. Bajkowski, The 1/2 subharmonic resonance and its transition to chaotic motion in a nonlinear oscillator, to appear in Int. J. Non-linear Mech.

[22] W. Szemplinska-Stupnika, Secondary resonances and an approximate model of transition to chaotic motion in nonlinear oscillators, J. Sound and Vibrations /to appear/.

[23] Tondl A. Nonlinear vibrations of mechanical systems Izd. Mir. Moskva, 1973.

[24] W. Y. Tseng and J. Dugundji, Nonlinear vibration of a buckled beam under harmonic excitation, J. Appl. Mech. 38, 1971, 467-476.

[25] Y. Tsuda, J. Inoue, H. Takamura and A. Sueoka, On the 1/2-the subharmonic vibrations of non-linear vibrating system with a hard Duffing type restoring characteristic, Bull. of JSME 27, 228, 1280-1287, June 1984.

[26] Y. Ueda, Randomly transition phenomena in the system governed by Duffing's equation, J. Statistical Physics, 20, 2, 181-196, 1979.

[27] Y. Ueda, Explosion of strange attractors exhibited by Duffing's equation, Annals of the New York Academy of Sciences, 357, 422-423, 1980.

[28] Y. Ueda, Steady motions exhibited by Duffing's equation: a picture book of regular and chaotic motions, New Approaches to Nonlinear Problems in Dynamics, SIAM, Philadelphia 1980.

[29] Y. Ueda, Chaotically transitional phenomena in the forced negative-resistance oscillator, IEEE Transaction on Circuits and Systems, 28, 217-223, 1981.

LOCAL TECHNIQUES IN BIFURCATION THEORY AND NONLINEAR DYNAMICS

G. Iooss
Université de Nice, Nice Cedex, France

1. ELEMENTARY STEADY BIFURCATIONS

1.1 First example

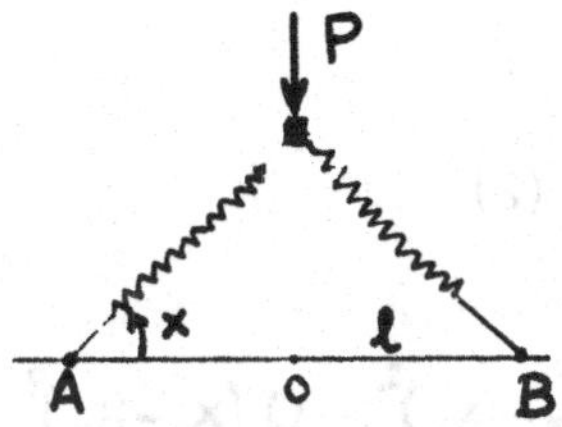

Figure 1

Let us consider a mass m fastened to two identical springs of constant k and natural length $\ell/\cos x_0$.

These springs are fixed respectively in A and B with $AB = 2\ell$. The mass m can move on oy , the symmetry axis of the system. There is a viscous friction on the axis, with constant K and a force P acts on the mass in the direction of the axis. P is positive when it is as on Figure 1. The equation which governs the movement of the mass is then as follows :

$$(1) \quad m\ell(\mathrm{tg}\,x)'' + K\ell(\mathrm{tg}\,x)' - 2k\ell\left[(\cos x_0)^{-1} - (\cos x)^{-1}\right]\sin x + P = 0.$$

Let us note $\mu = P/2k\ell$, then equilibria are given by $x = x_e$ with $\mu = \mu(x_e)$ such that :

(2) $$\mu - \sin x_e \left[(\cos x_o)^{-1} - (\cos x_e)^{-1}\right] = 0 .$$

The graph of (2) is drawn on Figure 2 .

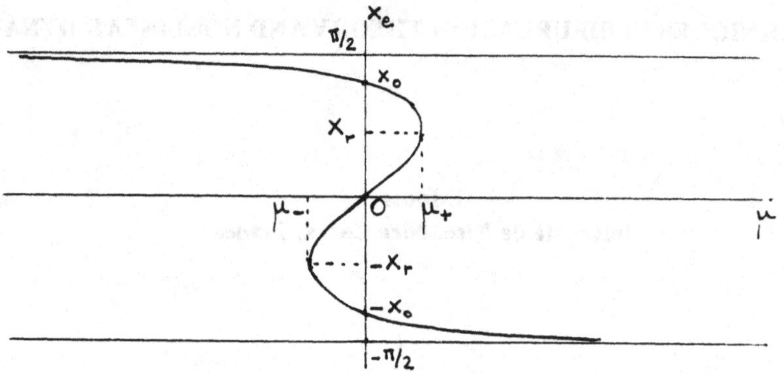

Figure 2

Let us denote by x_r the angle such that

$$\cos x_r = (\cos x_o)^{1/3} \quad , \quad x_r \in (0, \pi/2),$$

then we observe that

$$\mu'(x_r) = 0 ,$$

and for x close to x_r we have, in expanding (2) :

(3) $$\mu - \mu(x_r) = -\frac{3}{2} \frac{\left[1-(\cos x_o)^{2/3}\right]^{1/2}}{\cos x_o} (x_e - x_r)^2 + O(x_e - x_r)^3 .$$

For μ in a neighborhood of $\mu(x_r)$ we can then assert that for $\mu > \mu(x_r)$ there is no equilibrium position x_e close to x_r , while for $\mu < \mu(x_r)$ there are two such equilibria. We shall see later that they have opposite stabilities. The phenomenon which occurs at $\mu_+ = \mu(x_r)$ is called a saddle-node bifurcation.

REMARK . Physically we see that when μ (i.e. P) increases the angle x_e jumps from x_r to some negative angle and tends towards $-\pi/2$. Coming backwards by decreasing μ , the angle x_e increases until $-x_r$ which is reached for $\mu_- = -\mu_+$, then x_e jumps to some positive angle and tends towards $\pi/2$. We obtain an hysteresis phenomenon.

More generally let us observe that the saddle-node bifurcation will occur in this one-dimensional context if we have one equation of the form

(4) $$F(\mu, x) = 0 ,$$

where F is as many differentiable as we wish and when there is a solution (μ_r, x_r) such that

(5) $$F'_x(\mu_r, x_r) = 0 , \quad F'_\mu(\mu_r, x_r) \neq 0 , \quad F''_{x^2}(\mu_r, x_r) \neq 0 .$$

In fact, conditions (5) lead to a local expression for the solution $\mu(x)$ of (4) :

(6) $$\mu(x) = \mu_r - \frac{F''_{x^2}(\mu_r, x_r)}{2\, F'_\mu(\mu_r, x_r)} (x - x_r)^2 + O(x - x_r)^3$$

as it can be proved easily, using an identification process of successive powers of $x - x_r$ in the Taylor expansion of (4) . We shall see below in I.5 how to generalize conditions (5) and the computation in more than one dimension.

1.2 Second Example

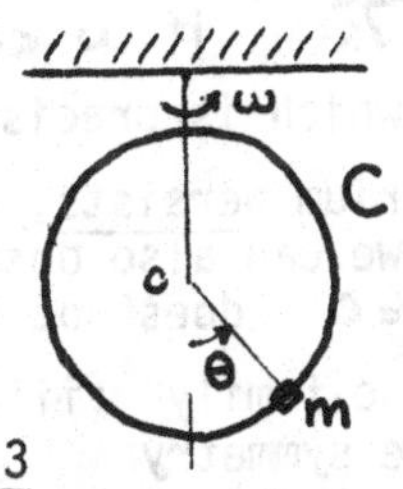

Figure 3

Let us consider a hoop C of radius R , rotating uniformly with a rotation rate ω about a vertical diameter oz . A mass m slides along C with a viscous friction, of constant k, and in presence of gravity. The laws of mechanics give us the equation of the movement :

(7) $$\ddot{\theta} + k\dot{\theta} + g/_R \sin\theta - \omega^2 \sin\theta \cos\theta = 0 .$$

Equilibrium positions are given by the solutions θ_e of

(8) $$\sin\theta_e \left(g/_R - \omega^2 \cos\theta_e \right) = 0.$$

Hence, the possible equilibria are :

i) $\theta_e = 0$, ii) $\theta_e = \pi$ and, if $\omega > \sqrt{g/R}$

iii) $\theta = \pm\theta_e$ such that $\cos\theta_e = g/\omega^2 R$,

$0 \leq \theta_e < \pi/2$. We give on Figure 4 the graph of equilibria in function of the parameter ω .

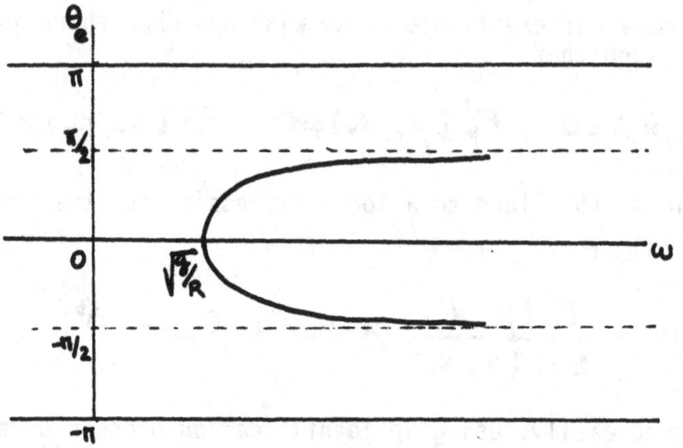

Figure 4

We observe here a new phenomenon occuring for $\omega = \sqrt{g/R}$. At this point we have the family (i) of equilibria given by $\theta_e = 0$ which " crosses " the family (iii) , making a " pitchfork " . This is named a " pitchfork bifurcation " . For ω close to $\sqrt{g/R}$, if $\omega < \sqrt{g/R}$ there is only one equilibrium close to θ_e which is precisely 0 , while for $\omega > \sqrt{g/R}$ the previous equilibrium persists, and two others appear in the neighborhood of 0 . Now we can also observe the following trivial fact here : the solution $\theta_e = 0$ does not break the symmetry $\theta \longmapsto -\theta$ of the problem, while the family (iii) breaks the symmetry since $\theta_e \neq 0$. Moreover the symmetry $\theta \longmapsto -\theta$ exchanges the two non symmetric equilibria. We shall see later that the stabilities of the symmetric and the non-symmetric branches are opposite.

If we write (8) like an equation of the form

$$F(\theta_e, \omega) = 0, \tag{9}$$

we see that F is odd in θ_e :

$$F(-\theta_e, \omega) = -F(\theta_e, \omega), \tag{10}$$

and we have

$$(11)\quad \begin{cases} F(0,\omega)=0 \quad , & F'_\theta(0,\sqrt{g/R})=0 \,, \quad F''_{\theta\omega}(0,\sqrt{g/R})\neq 0 \,, \\ F''_{\theta^2}(0,\omega)=0 \quad , & F'''_{\theta^3}(0,\sqrt{g/R})\neq 0 \,. \end{cases}$$

We shall see below in 1.6 how to generalize conditions (11) and to compute pitchfork bifurcations in more than one dimension. Here we have

$F''_{\theta\omega} = -2\sqrt{g/R}$, $F'''_{\theta^3} = 3g/R$, and

$$\theta_e = \pm \left[-6\left(\omega-\sqrt{g/R}\right) F''_{\theta\omega} / F'''_{\theta^3} \right]^{1/2} + O\left(\omega-\sqrt{g/R}\right)^{3/2},$$

and the " bifurcated branch " has an amplitude proportional to

$$\left(\omega - \sqrt{g/R}\right)^{1/2} \quad .$$

1.3 <u>Implicit function theorem in finite dimensions</u>

The idea is to solve with respect to x in $\mathbb{R}^n$ an equation of the form

$$(12)\qquad F(\mu, x) = 0 \qquad \text{in } \mathbb{R}^n \quad ,$$

where μ is a real parameter, knowing that

$$(13)\qquad F(\mu_0, x_0) = 0 \quad .$$

We introduce the Jacobian matrix

$$(14)\qquad A_0 = D_x F(\mu_0, x_0)$$

such that in the canonical basis of $\mathbb{R}^n$:

$$A_0 = \left(\frac{\partial F_i}{\partial x_j} \right)_{i,j=1,\dots n} \quad .$$

Then we have the following classical result, called <u>the Implicit function theorem</u> :

> If F is k times continuously differentiable, $k \geq 1$, in a neighborhood of (μ_0, x_0) , and satisfies (13) and A_0 invertible, then there exists in a neighborhood of (μ_0, x_0) a

unique solution $(\mu, x(\mu))$ of (12), k times continuously differentiable, such that $x(\mu_0) = x_0$.

Practically we often need to compute the Taylor expansion of $x(\mu)$:

$$x(\mu) = \sum_{p=c}^{k} (\mu-\mu_0)^p x + o(\mu-\mu_0)^k \tag{15}$$

Due to the uniqueness of the solution it is then sufficient to identify powers of $\mu-\mu_0$ into the expansion of

$$F[\mu, x(\mu)] = 0, \tag{16}$$

in the neighborhood of (μ_0, x_0). Let us introduce $\nu = \mu-\mu_0$, $y = x-x_0$, then, the Taylor expansion of F is as follows :

$$F(\mu, x) = \sum_{p+q=1}^{k} \nu^p F_{pq}[y^{(q)}] + o(|\nu| + \|y\|)^k, \tag{17}$$

where F_{pq} is q-linear symmetric in its argument, $y^{(q)}$ standing for the repetition of q identical arguments y. This means that each component of $F_{pq}[y^{(q)}]$ is an homogeneous polynomial of degree q of the components y_j of y. In fact we have

$$F_{pq} = \frac{1}{p!q!} \frac{\partial^{p+q} F(\mu_0, x_0)}{\partial \mu^p \partial x^q} \tag{18}$$

Equation (16) leads to a hierarchy of equations :

$$\begin{aligned} &A_0 x_1 + F_{10} = 0 \qquad , A_0 \equiv F_{01}, \\ &A_0 x_2 + F_{20} + F_{11} x_1 + F_{02}[x_1^{(2)}] = 0, \\ &\ldots\ldots\ldots \end{aligned} \tag{19}$$

Hence we deduce

$$\begin{aligned} x_1 &= - A_0^{-1} F_{10}, \\ x_2 &= - A_0^{-1} F_{20} - F_{11}(A_0^{-1} F_{10}) + F_{02}[(A_0^{-1} F_{10})^{(2)}], \\ &\ldots\ldots\ldots \end{aligned} \tag{20}$$

1.4 Fredholm alternative

We noticed that the computation of the expansion (15) relies upon the fact that A_0^{-1} exists, i.e. 0 is not an eigenvalue of A_0 in I.3. We can now observe that in the two examples treated in I.1. and I.2, x is one dimensional and in both examples

$$A_0 = 0$$

since it is $F_x'(\mu_r, x_r)$ in (5) and $F_\theta'(0, \sqrt{g/R})$ in (11). We now want to generalize these examples, so it is a motivation to study the resolution of the linear equation :

$$A_0 x = y \tag{21}$$

where y is given in $\mathbb{C}^n$, and where we assume that 0 is an eigenvalue of the linear operator A_0 in $\mathbb{C}^n$.

The basic result is that (21) has solutions if and only if y is orthogonal to the kernel of the adjoint operator A_0^*. Then x is determined up to an element of the kernel of A_0. We write this property as follows : $(\mathrm{Im}(A_0) = \text{range of } A_0)'$

$$\mathrm{Im}(A_0) \oplus \mathrm{Ker}(A_0^*) = \mathbb{C}^n \tag{22}$$

where the decomposition is orthogonal, and where A_0^* is defined by

$$(A_0 x \mid x') = (x \mid A_0^* x') \tag{23}$$

for any x, x' in $\mathbb{C}^n$, $(.|.)$ denoting a scalar product.

REMARK . When 0 is a semi-simple eigenvalue of A_0, i.e. when $A_0^2 x = 0$ is equivalent to $A_0 x = 0$, then

$$\mathrm{Im}(A_0) \oplus \mathrm{Ker}(A_0) = \mathbb{C}^n \tag{24}$$

holds too.

1.5. Saddle-node bifurcation in $\mathbb{R}^n$

We want to solve in $\mathbb{R}^n$

$$F(\mu, x) = 0 \quad , x \in \mathbb{R}^n$$

with the following assumptions :

H.1. $F(0,0) = 0$, μ close to 0 in $\mathbb{R}$,

x close to 0 in $\mathbb{R}^n$;

H.2. $A_0 = D_x F(0,0)$ has 0 as a simple eigenvalue.

The generalization of the second and third conditions in (5) needs the introduction of vectors ξ_0 and ξ_0^* such that

(25) $$A_0 \xi_0 = 0 \, , \quad A_0^* \xi_0^* = 0 \, , \quad (\xi_0 | \xi_0^*) = 1 \, .$$

Then we make the additional assumption :

H.3. $$(F_{10} | \xi_0^*) \neq 0 \quad , \quad (F_{02} [\xi_0^{(2)}] | \xi_0^*) \neq 0 ,$$

where we use the notations of (17) for the Taylor expansion of F in the neighborhood of 0 in $\mathbb{R}^{n+1}$.

The idea is to parameterize the branch of solutions in the following way :

(26) $$\begin{cases} x(\varepsilon) = \sum_{p=1}^{k} \varepsilon^p x_p + o(\varepsilon^k) \\ \mu(\varepsilon) = \sum_{p=1}^{k} \varepsilon^p \mu_p + o(\varepsilon^k) , \end{cases}$$

and to identify successive powers of ε in the expansion of

(27) $$F[\mu(\varepsilon), x(\varepsilon)] \equiv 0 \, .$$

The parameterization (26) is not unique. We may impose some normalization to fix it. Now, the order ε in (27) leads to

(28) $$\mu_1 F_{10} + A_0 x_1 = 0 \quad , \text{ since } \quad F_{01} = A_0 \, .$$

This equation has the form (21) . Since $(F_{10} | \xi_0^*) \neq 0$, due to H.3 , we obtain :

(29) $$\mu_1 = 0 \, , \quad x_1 = \alpha \xi_0 \, .$$

The normalization

(30) $$(x_1 | \xi_0^*) = 1 \quad , \quad (x_p | \xi_0^*) = 0 \quad , p \geq 2$$

leads to

(31) $$\alpha = 1 .$$

The order ε^2 in (27) gives us :

(32) $$\mu_2 F_{10} + A_0 x_2 + F_{02}[x_1^{(2)}] = 0$$

where the compatibility condition gives :

(33) $$\mu_2 = - \frac{(F_{02}[\xi_0^{(2)}] \mid \xi_0^*)}{(F_{10} \mid \xi^*)}$$

By H.3 we have $\mu_2 \neq 0$ (see Figure 5) . Moreover, we obtain x_2 , uniquely determined thanks to the normalization (30) . At higher orders, we obtain an equation of the form :

(34) $$\mu_p F_{10} + A_0 x_p + H(x_1, \dots, x_{p-1}, \mu_2, \dots, \mu_{p-1}) = 0 ,$$

where H is a known function of its arguments. Then we find μ_p by the compatibility condition and a unique x_p thanks to (30) .

REMARK . If F is analytic (its Taylor expansion converges in a ball $|\mu| + |x| < \rho$), then one can prove that the series (26) converge in some ball $|\mu| + |x| < \rho_1$

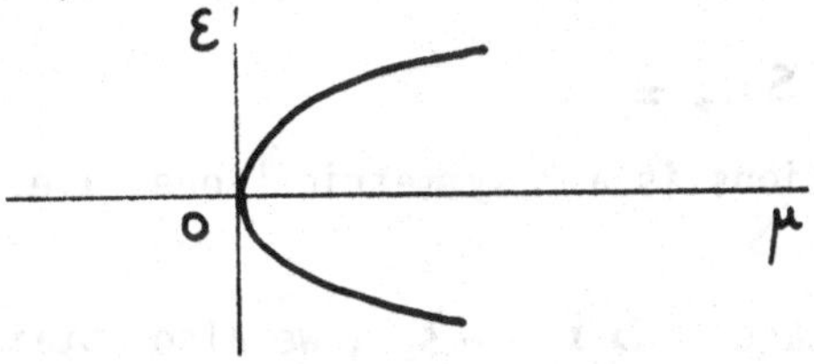

Figure 5 . Bifurcation diagram for a saddle-node bifurcation

1.6 Pitchfork bifurcation in $\mathbb{R}^n$

We want to generalize example 1.2 , i.e. we want to solve in $\mathbb{R}^n$, in a neighborhood of 0 :

(35) $$F(\mu, x) = 0 \quad , \quad x \in \mathbb{R}^n$$

with the assumptions H.1 and H.2 as in 1.5. Here we assume that there is a symmetry operator S , i.e.

(36) $$S^2 = Id \quad , \quad S \neq Id$$

such that F and S commute :

(37) $$F(\mu, Sx) = S F(\mu, x)$$

REMARK . In example 1.2 , S is the symmetry $\theta \longmapsto -\theta$, and it is obvious that the left hand side of (8) is odd in θ

In differentiating (37) at the origin, we see immediately :

(38) $$F_{pq}[(Sx)^{(q)}] = S F_{pq}[x^{(q)}] ,$$

hence

F_{po} is invariant under S , and

A_o commutes with S .

As a consequence, $S\xi_o$ is also an eigenvector belonging to the eigenvalue 0 for A_o , and since $S^2 = Id$ there are only two possibilities :

either $S\xi_o = \xi_o$, or $S\xi_o = -\xi_o$.

(i) In the first case $S\xi_o = \xi_o$, we also obtain $S^*\xi_o^* = \xi_o^*$ and it is easy to check (exercise left to the reader) that we can compute the expansion (26) in the same way as in 1.5, to obtain at each step x_p such that

$$Sx_p = x_p ,$$

hence the branch of solutions is a " symmetric " one, i.e. it is invariant under the symmetry S

(ii) In the second case $S\xi_o = -\xi_o$, we also obtain $S^*\xi_o = -\xi_o^*$, and we easily check that :

$$(F_{10} \mid \xi_o^*) = (S F_{10} \mid \xi_o^*) = (F_{10} \mid S^*\xi_o^*) = -(F_{10} \mid \xi_o^*)$$

$$(F_{02}[\xi_o^{(2)}] \mid \xi_o^*) = (F_{02}[(S\xi_o)^{(2)}] \mid \xi_o^*) = (F_{02}[\xi_o^{(2)}] \mid S^*\xi_o^*) = -(F_{02}[\xi_o^{(2)}] \mid \xi_o^*)$$

so the two coefficients in H.3 are zero.

Let us look for solutions parameterized in the same way as (26), and identify powers of ε in (27). We obtain at the order ε:

(39) $$\mu_1 F_{10} + A_0 x_1 = 0 \quad ,$$

where $(F_{10} \mid \xi_0^*) = 0$. Hence we obtain

(40) $$x_1 = \alpha \xi_0 - \tilde{A}_0^{-1} \mu_1 F_{10} \quad ,$$

where $\tilde{A}_0^{-1}$ is the inverse of the restriction of A_0 acting in $\mathrm{Im}(A_0)$, and α and μ_1 are still arbitrary since the normalization which fixes the parameterization is not yet fixed. In fact, one can show that there are two families of solutions, one which is symmetric and persistent when μ crosses 0, and another which breaks the symmetry and occurs only on one side of $\mu = 0$. Let us start with

(41) $$\alpha = 0 \quad , \quad \mu_1 = 1 \quad ,$$

then order ε^2 leads to:

(42) $$F_{10} \mu_2 + A_0 x_2 + F_{20} + F_{11} x_1 + F_{02} [x_1^{(2)}] = 0 \, .$$

We can impose

(43) $$\mu_p = 0 \quad , \quad p \geq 2 \quad ,$$

and solve (42), since

$$F_{20} + F_{11} x_1 + F_{02} [x_1^{(2)}]$$

is invariant under S, hence orthogonal to ξ_0^*. We can compute in the same way any x_p, $p \geq 2$ and then obtain $X(\varepsilon)$ invariant under S, with $\mu = \varepsilon$. This is the persistent symmetric branch of solutions of (35). Let us now consider the case when

(44) $$\alpha = 1 \quad , \quad \mu_1 = 0$$

with a normalisation such that

$$(x_p \mid \xi_0^*) = 0 \quad , \quad p \geq 2 \quad .$$

Then, order ε^2 leads to:

(45) $$F_{10} \mu_2 + A_0 x_2 + F_{02} [x_1^{(2)}] = 0 \quad ,$$

hence for the same reason as above, it is solvable with respect to x_2, with no condition on μ_2. Now the order ε^3 gives us :

(46) $$\mu_3 F_{10} + A_0 x_3 + \mu_2 F_{11} x_1 + 2 F_{02}[x_1, x_2] + F_{03}[x_1^{(3)}] = 0 .$$

The compatibility condition gives μ_2 provided that the following assumption holds :

(47) $$H.'3 : \Lambda_1 = (F_{11}\xi_0 | \xi_0^*) - 2 (F_{02}(\xi_0, \tilde{A}_0^{-1} F_{10}) | \xi_0^*) \neq 0 .$$

Then we have

(48) $$\mu_2 = \Lambda_1^{-1}([2 F_{02}(\xi_0, \tilde{A}_0^{-1} F_{02}[\xi_0^{(2)}]) - F_{03}[\xi_0^{(3)}]] | \xi_0^*),$$

and we can impose

(49) $$\mu_{2p+1} = 0 \quad , p \geq 1 .$$

The computation of μ_{2p}, $p \geq 2$ is analogous to the computtaion of μ_2, and can be done, thanks to H!3 . By this way we observe the following property for the series (26) :

(50) $$\begin{cases} S x_{2k+1} = - x_{2k+1} , & S x_{2k} = x_{2k}) \\ \mu(\varepsilon) = \sum_{k=1}^{K} \varepsilon^{2k} \mu_{2k} + o(\varepsilon^{2K}) \end{cases}$$

In fact, we have

(51) $$S x(\varepsilon) = x(-\varepsilon) , \quad \mu(\varepsilon) = \mu(-\varepsilon)$$

as this property might be proved for the full function $x(\varepsilon)$ (not only for the finite Taylor expansion). This shows the symmetry breaking property and it is summed up at Figure 6 .

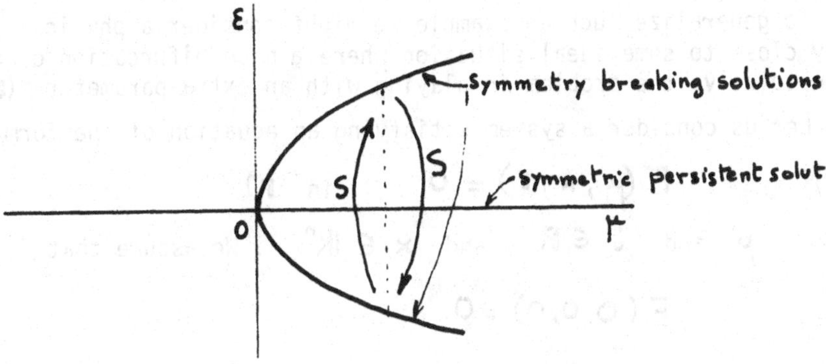

Figure 6 . Pitchfork bifurcation diagram - ε is the coefficient of the eigenvector ξ_0 into the solution.

1.7 Elementary imperfection theory

Physical problems are often posed into a too much idealistic way to be close to reality. Let us go back to the example 1.2 and assume that in O there is a weak spring acting on Om with a small moment $mR^2\delta$ (so, without gravity nor ω, the mass m should rotate along C). Now, instead of (8) we have

(52) $$\delta = \sin\theta_e\,(g/R - \omega^2\cos\theta_e) .$$

If we expands the right hand side for ω close to $\sqrt{g/R}$ and θ_e close to O we obtain :

(53) $$\delta = 2\sqrt{g/R}\;\theta\left[(\sqrt{g/R} - \omega) + 1/4\sqrt{g/R}\;\theta^2 + \cdots\right] .$$

It is easy to study the behavior of the graph of (52) in the plane (ω, θ_e). This is summed up on Figure 7 .

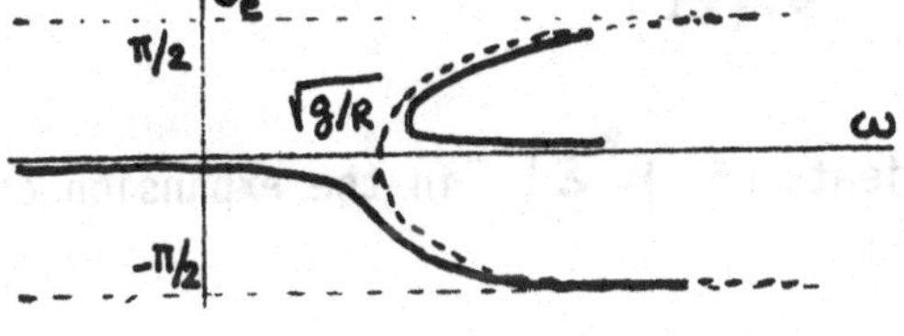

Figure 7 . (case $\delta > 0$) graph of (52) for $\delta < g/R$.

We see that the bifurcation is broken and that (53) describes the behavior of the graph of (52) in the neighborhood of the bifurcation point.

To generalize such an example we might consider a physical problem very close to some ideal situation where a nice bifurcation occurs, and then to solve the problem in playing with an extra-parameter (δ) .

Let us consider a system satisfying an equation of the form

(54) $$F(\mu, x, \delta) = 0 \quad \text{in } \mathbb{R}^n$$

where μ and $\delta \in \mathbb{R}$ and $x \in \mathbb{R}^n$. We assume that

(55) $$F(0,0,0) = 0$$

and that for $\delta = 0$ we have a bifurcation problem :

(56) $A_0 = D_x F(0,0,0)$ has 0 as a simple eigenvalue.

We denote as before by ξ_0 and ξ_0^* the eigenvectors of A_0 and A_0^* belonging to 0 , and we assume moreover :

(57) H.4 $(D_\delta F(0,0,0) \mid \xi_0^*) \neq 0.$

Let us expand F in a neighborhood of 0 of $\mathbb{R}^{n+2}$ as follows :

(58) $$F(\mu, x, \delta) = \sum_{p+q+r \geq 1}^{k} \mu^p \delta^q F_{pqr}[x^{(r)}] + o\left(|\mu| + |\delta| + \|x\|\right)^k ,$$

where F_{pqr} is r-linear symmetric in its arguments. The idea is to look for a branch of solutions of the form

(59) $$\begin{cases} x(\mu, \varepsilon) = \sum_{p+q \geq 1} \mu^p \varepsilon^q x_{pq} \\ \delta(\mu, \varepsilon) = \sum_{p+q \geq 1} \mu^p \varepsilon^q \delta_{pq} , \end{cases}$$

and to identify coefficients of $\mu^p \varepsilon^q$ in the expansion of (54) at the neighborhood of zero.

1.7.1 Perturbation of a saddle-node bifurcation

Let us assume that for $\delta = 0$, the assumption H.3 holds :

H.3 $(F_{100} \mid \xi_0^*) \neq 0 \quad , \quad (F_{002}[\xi_0^{(2)}] \mid \xi_0^*) \neq 0 .$

The order 1 in (54) leads to the following two equations :

(60) $$F_{100} + \delta_{10} F_{010} + A_0 x_{10} = 0 ,$$

(61) $$\delta_{01} F_{010} + A_0 x_{01} = 0$$

This gives

(62) $$\delta_{10} = - (F_{100} \mid \xi_0^*) / (F_{010} \mid \xi_0^*) \quad , \delta_{01} = 0 ,$$

and tne normalization

$$(x_{pq} \mid \xi_0^*) = 0 \quad \text{for } (p,q) \neq (0,1) \; , (x_{01} \mid \xi_0^*) = 1 ,$$

leads to

(63) $$x_{01} = \xi_0 \; , \quad x_{10} = - \tilde{A}_0^{-1} (F_{100} + \delta_{10} F_{010}) .$$

We may notice that

$$(F_{010} \mid \xi_0^*) \neq 0 \quad , \text{ due to H.4} \; .$$

The order 2 gives 3 equations :

(64) $$\delta_{20} F_{010} + A_0 x_{20} + F_{200} + \delta_{10} F_{110} + (\delta_{10})^2 F_{020} + \\ + F_{101} x_{10} + \delta_{10} F_{011} x_{10} + F_{002} [x_{10}^{(2)}] = 0 ,$$

(65) $$\delta_{11} F_{010} + A_0 x_{11} + F_{101} x_{01} + \delta_{10} F_{011} x_{01} + 2 F_{002} [x_{01}, x_{10}] = 0,$$

(66) $$\delta_{02} F_{010} + A_0 x_{02} + F_{002} [x_{01}^{(2)}] = 0$$

The compatibility conditions give δ_{20} , δ_{11} , δ_{02} and the normalization leads to unique x_{20} , x_{11} , x_{02} . We see that assumption H.3 gives $\delta_{02} \neq 0$ since

(67) $$\delta_{02} = - (F_{002} [\xi_0^{(2)}] \mid \xi_0^*) / (F_{010} \mid \xi_0^*) .$$

If we consider the graph of

$$\delta = \delta(\mu, \varepsilon) = \delta_{10} \mu + \delta_{20} \mu^2 + \delta_{11} \mu \varepsilon + \delta_{02} \varepsilon^2 + \ldots .$$

in the plane (μ,ε) then its looks like indicated at Figure 8 . It is just a regular perturbation of the graph of Figure 5 .

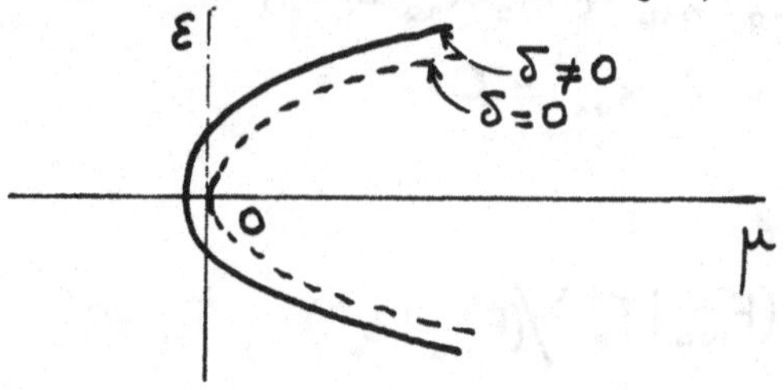

Figure 8 . Imperfect saddle-node bifurcation. We then obtain again a saddle-node bifurcation, slightly shifted in the μ direction.

1.7.2 Perturbation of a pitchfork bifurcation

Let us assume that for $\delta = 0$, the system is invariant under a symmetry S , as in 1.6 :

(68) $$F(\mu, Sx, 0) = S\,F(\mu, x, 0) .$$

We assume here again that $S\xi_0 = -\xi_0$ and that H.4 holds :

(69) $$(F_{010} \mid \xi_0^*) \neq 0 ,$$

and here the coefficients occuring in H.3 are both zero (as in 1.6). The order 1 in (μ, ε) into (54) here gives again (60) (61) , but now :

(70) $$\delta_{10} = \delta_{01} = 0 \quad , \quad x_{01} = \xi_0 \quad , \quad x_{10} = -\tilde{A}_0^{-1} F_{100} .$$

The order 2 now leads to (64) , (65) , (66) with $\delta_{10} = 0$. The solution may be written as follows :

(71) $$x_{20} = -\tilde{A}_0^{-1}\left\{F_{200} + F_{101}x_{10} + F_{002}[x_{10}^{(2)}]\right\} \quad , \quad \delta_{20} = 0 ,$$

(72) $$x_{11} = -\tilde{A}_0^{-1}\left\{F_{101}\xi_0 + 2F_{002}[\xi_0, x_{10}] + \delta_{11}F_{010}\right\} ,$$

(73) $$\delta_{11} = -\Lambda_1/(F_{010} \mid \xi_0^*) \quad \text{with}$$

$$\Lambda_1 = (F_{101}\xi_0 \mid \xi_0^*) - 2\left(F_{002}[\xi_0, \tilde{A}_0^{-1}F_{100}] \mid \xi_0^*\right) ,$$

(74) $\delta_{02} = 0 \quad , \quad x_{02} = -\tilde{A}_0^{-1} F_{002}[\xi_0^{(2)}]$.

At the order 3 we obtain 4 equations, and it is easy to check that we obtain :

(75) $\delta_{30} = \delta_{12} = 0$,

(76) $\delta_{03} = -\left(2F_{002}[\xi_0, x_{02}] + F_{003}[\xi_0^{(3)}] \mid \xi_0^*\right) / (F_{010} \mid \xi_0^*)$,

and more generally we have

$$\delta_{pq} = 0 \qquad \text{for} \quad q \ \underline{\text{even}}.$$

Finally the graph of

(77) $\delta = \delta(\mu, \varepsilon) = \delta_{11}\mu\varepsilon + \delta_{03}\varepsilon^3 + \delta_{21}\mu^2\varepsilon + \dots$

looks locally like the graph of

$$\delta = (\delta_{11}\mu + \delta_{03}\varepsilon^2)\varepsilon$$

since $\mu^2\varepsilon$ is $o(\mu\varepsilon)$. We recover again (48) for $\delta = 0$. The situation is summed up on Figure 9 .

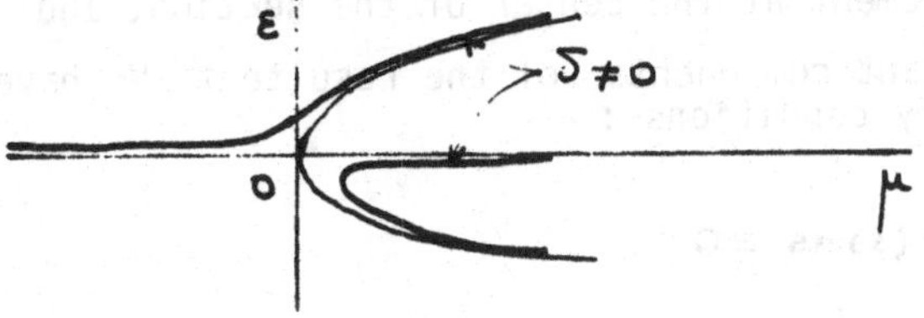

Figure 9

We observe onto the expansion of the branch of solutions that $S x_{pq} = (-1)^q x_{pq}$ as in (50) , but the curve on Figure 9 is no longer symmetric with respect to the μ-axis, hence $\mu(-\varepsilon)$ is not equal to $\mu(\varepsilon)$, and solutions with positive ε or with negative ε are physically different.

1.8 Example - Buckling of an elastic beam

1.8.1. Modelization

We consider an unextensible beam of length ℓ . The point O is fixed and A may move along the x axis. A force P is applied in A along ox. The boundary conditions in O and A can be either clamped conditions or perfect knee-joint conditions. We only consider static equilibrium (Figure 10) and we assume that the curve describing the shape of the beam is a plane curve (lying in xy plane)

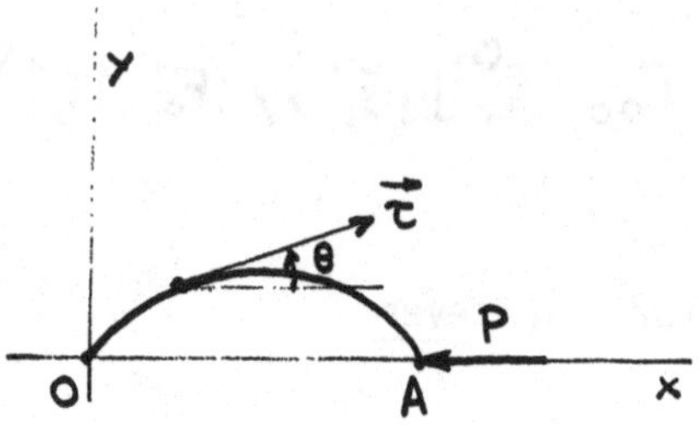

Figure 10

A classical study of curvilinear mechanics leads to the following equation :

(78) $$\frac{dM}{ds} + T\cos\theta + P\sin\theta = 0 ,$$

where s is the curvilinear abscissa $s \in (0,\ell)$, θ is the angle of the tangent $\vec{\tau}$ to the curve with the x axis, M is the z-component of the moment at the center of the section, and $(-P,T)$ are the (x,y) constant components of the resultent. We have in addition, due to boundary conditions :

(79) $$\int_0^\ell \sin\theta(s)\,ds = 0,$$

from which we deduce that

(80) $$M(\ell) - M(0) + T\int_0^\ell \cos\theta(s)\,ds = 0 .$$

The elastic behavior of the beam is ruled by :

(81) $$M = EI\frac{d\theta}{ds}$$

where I is the inertial moment of the section of the beam with respect to the z axis, E is the Young modulus of the material, $d\theta/ds$ is the curvature.

Let us now choose the simplest problem, which is when in O and in A we have perfect knee-joints. In this case $M(0) = M(\ell) = 0$, hence $T = 0$ and we arrive to the Euler-

Bernoulli problem.

Let us introduce

(82) $s = \ell x \ , \ \lambda = P\ell^2/EI \ , \ \theta(s) = Y(x),$

then the system becomes :

$$(83) \quad \begin{cases} \dfrac{d^2Y}{dx^2} + \lambda \sin Y = 0 \quad , x \in [0,1] , \\ Y'(0) = Y'(1) = 0 . \end{cases}$$

Since $(83)_1$ is just the pendulum equation, a nearly explicit solution can be derived in some standard way. It is not our aim here.

1.8.2. The Pitchfork bifurcation

We observe that (83) may be understood as an equation of the form

(84) $F(Y,\lambda) = 0 ,$

where Y lies in a space E of regular functions satisfying the boundary condition $(83)_2$. Let us choose

$$E = \{ Y \in \mathcal{C}^2[0,1] \ ; \ Y'(0) = Y'(1) = 0 \} .$$

So, in this example E is infinite dimensional, but we shall see how to handle this problem as in $\mathbb{R}^n$. We have $F(0,\lambda) = 0$ and

(85) $F(-Y,\lambda) = -F(Y,\lambda)$

shows that we have a symmetry (have very obvious). Let us consider the linearization of F about 0 :

(86) $A_\lambda Y = \dfrac{d^2Y}{dx^2} + \lambda Y \ , \ Y \in E .$

All the analysis we did in $\mathbb{R}^n$ relies upon the possibility to solve an equation :

(87) $A_\lambda Y = Z$

where Z is given in $\mathcal{C}^0[0,1]$ (the space where $F(Y,\lambda)$ lies) and where we look for Y in E .

It is then well known [4] that if $\lambda \neq k^2\pi^2$, $k \in \mathbb{N}$, then A_λ is invertible in (87), while if $\lambda = k^2\pi^2$ we have 0 as an eigenvalue of A_λ and a compatibility condition is needed on Z to be able to obtain Y.

Let us consider the lowest physical critical value $\lambda = \pi^2$. Then the eigenvector is

$$\xi_0 = \cos \pi x$$

and the compatibility condition reads

$$\int_0^1 Z(x)\cos(\pi x)\,dx = 0 .$$

Then there is a unique solution of $A_{\pi^2} Y = Z$, if we impose

$$\int_0^1 Y(x)\cos(\pi x)\,dx = 0,$$

since this condition determines α into

$$Y(x) = \alpha\cos(\pi x) + \frac{1}{\pi}\int_0^x \sin[\pi(x-s)]\,Z(s)\,ds .$$

Now all the method for computing the series (26) works as in $\mathbb{R}^n$. We pose

$$\text{(88)}\quad \begin{cases} Y = \sum_{p\geq 1} \varepsilon^p Y_p , \quad \int_0^1 Y_1\cos(\pi x)\,dx = 1/2 , \quad \int_0^1 Y_p\cos(\pi x)\,dx = 0, \; p \geq 2 \\ \lambda = \pi^2 + \sum_{p\geq 1} \varepsilon^p \lambda_p \end{cases}$$

and an exercise for the reader would be to check that :

$$\text{(89)}\quad \begin{cases} Y_1 = \xi_0 = \cos\pi x , \quad Y_2 = 0, \\ Y_3 = -\frac{1}{192}\cos 3\pi x, \ldots \\ \lambda_1 = 0, \quad \lambda_2 = \pi^2/8 , \quad \lambda_3 = 0, \ldots \end{cases}$$

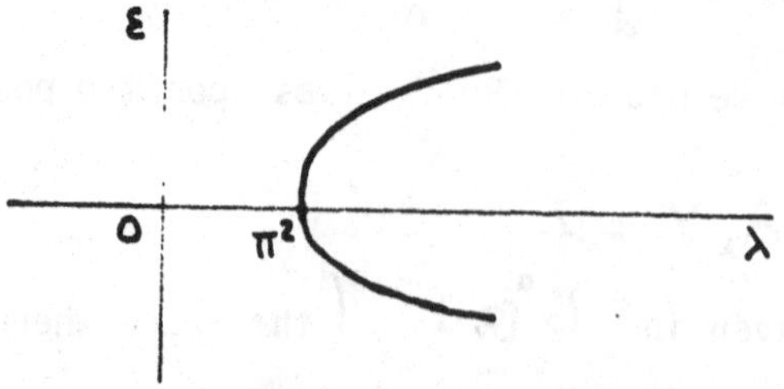

Figure 11 . Bifurcation diagram

1.8.3. Imperfect bifurcation

Let us assume that at the left end O , a little moment $M = -EI\delta/\ell$ is applied, where δ is small (Figure 12).

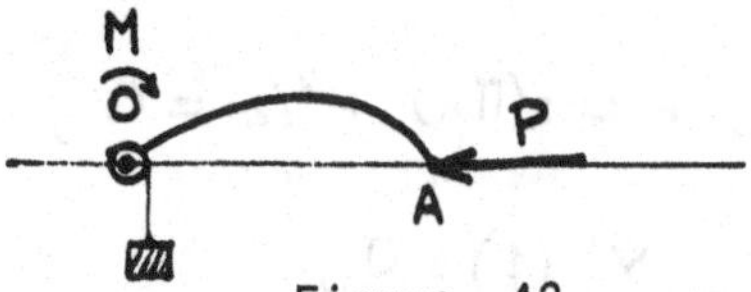

Figure 12

Now the boundary conditions in O and A become

(90) $$Y'(0) = \delta \quad , \quad Y'(1) = 0 ,$$

and the equilibrium equation is now :

(91) $$\frac{d^2 Y}{dx^2} + \lambda \sin Y + \delta \frac{\cos Y}{\int_0^1 \cos Y(s)\, ds} = 0 ,$$

as it results from (78) (80) (81) .

We might proceed as in 1.8.2., in substracting $\delta x - \delta/2\, x^2$ to Y in such a way that the boundary conditions are now homogeneous, and to be able to work into the space E defined in 1.8.2., for an equation of the form

$$F(Y, \lambda, \delta) = 0 .$$

In fact this is not necessary and the idea is to look for series of the form :

(92) $$\begin{cases} Y = \sum \mu^p \varepsilon^q Y_{pq} \quad , \quad \mu = \lambda - \pi^2 , \\ \delta = \sum \mu^p \varepsilon^q \delta_{pq} \end{cases}$$

where $Y_{pq} \in \mathcal{C}^2[0,1]$ and $Y'_{pq}(0) = \delta_{pq}$, $Y'_{pq}(1) = 0$.

Replacing into (91) and identifying coefficients of $\mu^p \varepsilon^q$, with a normalization such that

$$\int_0^1 Y_{01} \cos(\pi x)\, dx = \frac{1}{2} \quad , \quad \int_0^1 Y_{pq} \cos(\pi x)\, dx = 0$$

for $(p,q) \neq (0,1)$

we then obtain (the reader is kindly requested to check) :

$$\begin{cases} Y_{01} = \xi_0 = \cos \pi x \ , & Y_{10} = 0, \\ \delta_{11} = 1/2 \ , & \delta_{03} = -\pi^2/16 \end{cases} \qquad \text{(see (77))}$$

and Y_{11} is the solution of :

$$\begin{cases} \dfrac{d^2 Y_{11}}{dx^2} + \pi^2 Y_{11} + \cos(\pi x) + 1/2 = 0 \ , \\ Y'_{11}(0) = 1/2 \ , \quad Y'_{11}(1) = 0 \ , \end{cases}$$

hence :

$$Y_{11} = -\frac{1}{2\pi^2} + \frac{1}{2\pi}(1-x)\sin(\pi x) + \alpha \cos(\pi x) .$$

We have a graph in the (μ, ε) plane like at Figure 9 for $\delta < 0$ (for $\delta > 0$ make just a symmetry with respect to the μ axis !).

2. CENTER MANIFOLD AND AMPLITUDE EQUATIONS

2.1 Hyperbolic situation

2.1.1 Linear case.

Let us consider the following linear differential equation in $\mathbb{R}^n$:

(1) $$\frac{dx}{dt} = Lx \quad , \; x(t) \in \mathbb{R}^n .$$

It is well known that if x_0 denotes the initial value, then the solution of (1) may be written as :

(2) $$x(t) = e^{Lt} x(0)$$

Where e^{Lt} is a time dependent linear operator in $\mathbb{R}^n$, such that

(3) $$\begin{cases} e^{Lt_1} \cdot e^{Lt_2} = e^{L(t_1+t_2)} & \text{, for any } t_1 \text{ and } t_2 \text{ in } \mathbb{R} , \\ e^{L0} = Id \; , \quad \frac{d}{dt} e^{Lt} = L e^{Lt} = e^{Lt} L . \end{cases}$$

Examples :

i) in $\mathbb{R}^1$, $L = k \quad e^{Lt} \equiv e^{kt}$ (the usual one).

ii) in $\mathbb{R}^2$,

if $L = \begin{pmatrix} \lambda_1 & 0 \\ 0 & \lambda_2 \end{pmatrix}$, then $e^{Lt} = \begin{pmatrix} e^{\lambda_1 t} & 0 \\ 0 & e^{\lambda_2 t} \end{pmatrix}$,

if $L = \begin{pmatrix} \lambda & 1 \\ 0 & \lambda \end{pmatrix}$, then $e^{Lt} = \begin{pmatrix} e^{\lambda t} & te^{\lambda t} \\ 0 & e^{\lambda t} \end{pmatrix}$,

if $L = \begin{pmatrix} 0 & -\omega \\ \omega & 0 \end{pmatrix}$, then $e^{Lt} = \begin{pmatrix} \cos \omega t & -\sin \omega t \\ \sin \omega t & \cos \omega t \end{pmatrix}$.

There is a general way to write e^{Lt} by using the Jordan decomposition of L [8]

Let us assume that some eigenvalues of L have a strictly positive real part, while all the others have strictly negative real part (see figure 1) .

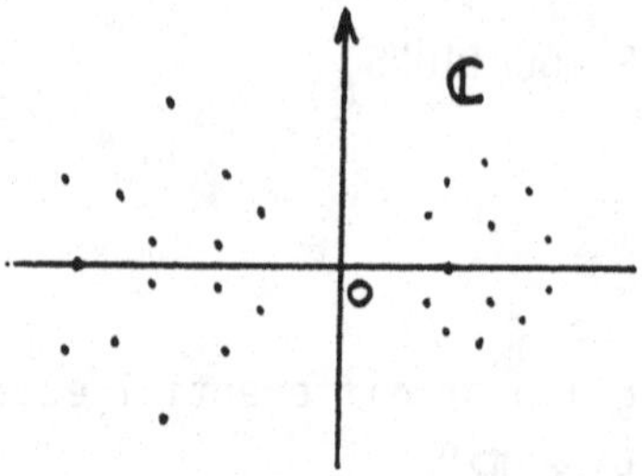

Figure 1. Spectrum of L (no eigenvalue on the imaginary axis)

Then the space $\mathbb{R}^n$ can be splitted into two subspaces E_+ and E_- which are invariant under L :

(4) $$LE_+ \subset E_+ \quad , \quad LE_- \subset E_- \quad , \quad \mathbb{R}^n = E_+ \oplus E_- \, ,$$

and the restrictions L_+ and L_- of L to E_+ and E_- have only the eigenvalues of strictly positive real part or strictly negative real part. Let us define P_+ and P_- the projections onto E_+ and E_-, which commute with the operator L. Then we can write the solution of (1) under the form :

(5) $$x(t) = e^{L_+ t} P_+ x_0 + e^{L_- t} P_- x_0 \, ,$$

since P_+ and P_- commute with e^{Lt}.

If we consider the part $e^{L_- t} P_- x_0$ of the right hand side of (5), where all the eigenvalues are such that

$$\mathrm{Re}\, \lambda_- < -\xi < 0 \, ,$$

then there exists a constant K such that

(6) $$|e^{L_- t} P_- x_0| \leq K e^{-\xi t} |P_- x_0| \text{ for } t \geq 0 \, .$$

In the same way we have

$$\mathrm{Re}\, \lambda_+ > \xi > 0$$

after choosing suitably ξ, and there exists K' such that :

(7) $$|e^{L_+ t} P_+ x_0| \geq K' e^{\xi t} |P_+ x_0| \quad , t \geq 0 \, .$$

As a consequence the dynamics is as indicated on figure 2

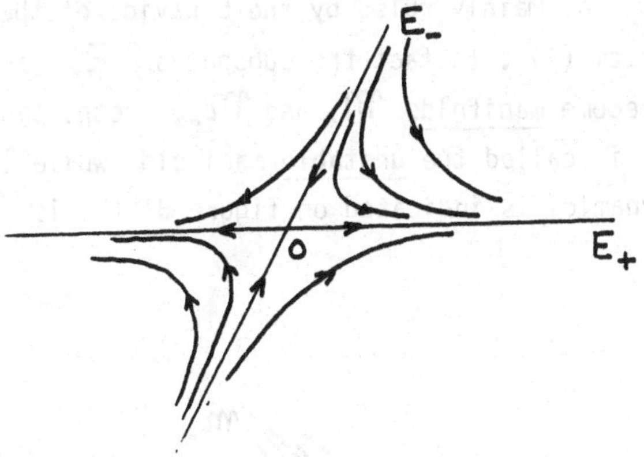

Figure 2.

Dynamics in the linear hyperbolic case.

We shall say that we have a saddle point in 0 . The only possibility to obtain $x(t) \to 0$ when $t \to \infty$ is to choose x_0 in E_- (then $x(t)$ stays in E_-.) .

2.1.2 . Stable and unstable manifolds

We consider now an evolution problem ruled by the following differential equation :

(3) $$\frac{dx}{dt} = F(x) \quad , \; x(t) \in \mathbb{R}^n .$$

We assume that

(9) $$F(0) = 0$$

i.e $x = 0$ is a steady solution of our system. Let us denote by

(10) $$L = D_x F(0)$$

the linear operator whose matrix is the Jacobian matrix

$$\left(\frac{\partial F_i}{\partial x_j} \right)_{i,j=1,\dots n}$$

We make on L the same assumption as above, then the dynamics in the neighborhood of 0, is mainly ruled by the behavior of the solution of the linearized system (1). In fact the subspaces E_+ and E_- of the linear situation become manifolds $\mathcal{M}_+$ and $\mathcal{M}_-$, resp. tangent to E_+ and E_-. $\mathcal{M}_+$ is called the unstable manifold while $\mathcal{M}_-$ is the stable one. The dynamics is indicated on figure 3 (0 is saddle point).

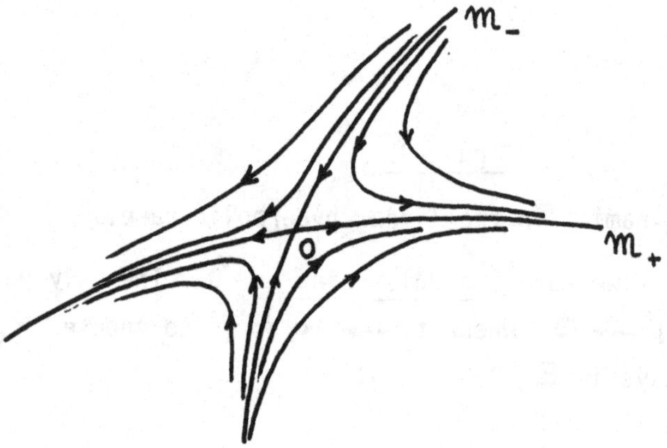

Figure 3. Dynamics in the neighborhood of 0 in the hyperbolic case.

A subcase is when all eigenvalues of L have a strictly negative real part. In this case, for any initial data close enough to 0, the solution $x(t)$ goes exponentially to 0 when t tends towards $+\infty$. We then say that 0 is a stable node.

2.1.3 . Structural stability of the hyperbolic situation.

If we consider a system which perturbs an hyperbolic situation, then we again obtain a saddle point like the unperturbed case. To see this let us write the new system on the form

(11) $$\frac{dx}{dt} = F(\mu, x) \quad , \mu \in \mathbb{R} \quad , x(t) \in \mathbb{R}^+ ,$$

and assume that $F(0,0) = 0$ and $D_x F(0,0) = L_0$ has no eigenvalue on the imaginary axis. First it is clear that for μ close to 0 there is a steady solution of (11)

(12) $$x = X_0(\mu) .$$

This comes from implicit function theorem (see I.3) since $F(0,0) = 0$ and $D_x F(0,0) = L_0$ is invertible. We can easily compute the Taylor expansion of (12) . Let us now introduce

(13) $$x = x_0(\mu) + y$$

then (11) becomes

(14) $$\frac{dy}{dt} = G(\mu, y) ,$$

with $G(\mu, 0) = 0$ and

(15) $$D_y G(\mu, 0) = L_\mu = L_0 + O(\mu) .$$

By the classical result on the continuity of eigenvalues with respect to parameter μ , we can assert that for μ close to 0 , the eigenvalues of L_μ are close to those of L_0 , hence there is no eigenvalue of L_μ on the imaginary axis for $|\mu| < \delta$. The dimensions of the invariant subspaces $E_+(\mu)$, $E_-(\mu)$ are constant and the stable and unstable manifolds $\mathcal{M}_-(\mu)$ and $\mathcal{M}_+(\mu)$ look the same as for $\mu = 0$. It can be proved that the topology of trajectories is exactly the same as for $\mu = 0$. We shall say that a saddle point is <u>structurally stable</u>

2.2 . Singular situations.

2.2.1 . Linear case.

Let us consider a system obeying a linear differential equation in $\mathbb{R}^n$

(16) $$\frac{dz}{dt} = Lz \quad , \quad z(t) \in \mathbb{R}^n$$

where we assume that the linear operator L has some eigenvalues on the imaginary axis and all the others with a nagative real part (figure 4) :

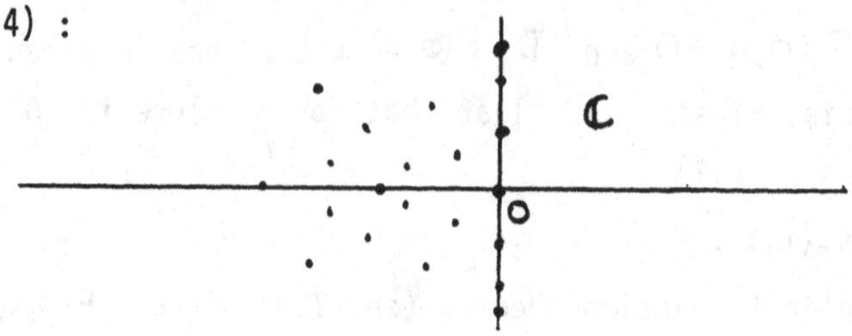

Figure 4 : Spectrum of L .

To understand the dynamics of the solutions of (16) , we may again split the space $\mathbb{R}^n$ as follows :

(17) $$\mathbb{R}^n = E_0 \oplus E_- ,$$

and we denote by L_0 and L_-. The restrictions of L to the invariant subspaces E_0 and E_- . We again obtain an estimate (6) for the contracting part $e^{L_- t}P_-$, but now the "central part" $e^{L_0 t}P_0$ has a special behavior. If L_0 is diagonalizable, then $e^{L_0 t}P_0$ is oscillating (constant if 0 is the only eigenvalue of L_0) , while if L_0 is not diagonalizable $e^{L_0 t}P_0$ behaves like a sum of polynomials of t, times oscillating functions (see Figure 5) .

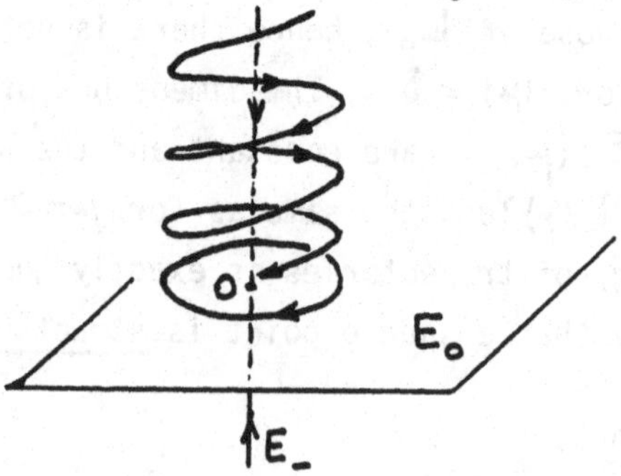

Figure 5. Example of the dynamics for the linear central case.

2.2.2. Center manifold

As in §2.1.2, let us consider the corresponding non-linear situation :

(18) $$\frac{dz}{dt} = F(z) \quad , \quad z(t) \in \mathbb{R}^n \, , \; F(0) = 0 \, ,$$

where we assume that the linear operator

(19) $$L = D_z F(0)$$

has the properties described at §2.2.1. Then the result is as follows (see [6] for references) :

there is a so-called "center manifold" $\mathcal{M}_0$ tangent to E_0, which can be expressed by :

(20) $$z = x + \varphi(x) \, ,$$

where $x \in E_0$, $\varphi(x) \in E_-$, $\varphi(0) = 0$, $D_x\varphi(0) = 0$, and such that :

i) $\mathcal{M}_0$ is locally invariant , i.e if the initial condition is on $\mathcal{M}_0$, then $z(t)$ stays on $\mathcal{M}_0$ provided that are stays in a a neighborhood of 0,

ii) $\mathcal{M}_0$ is locally attracting , i.e. for any initial condition, if $z(t)$ stays in same neighborhood of 0, then the distance between $z(t)$ and $\mathcal{M}_0$ tends to 0 when $t \to \infty$.

Remark . The center manifold is not unique in general (see examples in [6]) but its Taylor expansion at the origin is unique. This allows to compute explicitely this expansion, as we shall see below.

As a consequence of the attractivity of $\mathcal{M}_0$, the dynamics of the solutions of (18) in a neighborhood of 0 is then reduced to the study of the dynamics on $\mathcal{M}_0$. Hence we reduced the problem to a lower dimensional one, its dimension being the dimension of E_0. To illustrate this advantage let us consider the simplest example. Assume that 0 is a simple eigenvalue of L , and that all other eigenvalues have a negative real part. Let us denote by ξ_0 and ξ_0^* the eigenvectors such that (see §1.5)

$$L\xi_0 = 0 \quad , \quad L^*\xi_0^* = 0 \quad , \quad (\xi_0 | \xi_0^*) = 1 \,. \tag{21}$$

The center manifold takes a priori the following form

$$z = x\xi_0 + \varphi(x) \qquad \text{with} \quad x \in \mathbb{R} \,, \tag{22}$$

and $\varphi(x)$ orthogonal to ξ_0^* satisfies $\varphi(0) = \varphi'(0) = 0$.

The dynamics on the one dimensional center manifold is now ruled by the differential equation

$$\frac{dx}{dt} = f(x) \quad , \qquad f(0) = f'(0) = 0 \,. \tag{23}$$

Let us show how to explicitely find the expansions of φ and f .

For this purpose we need to introduce some notations (as in §1.3) :

$$\begin{cases} F(z) = Lz + \sum_{p \geq 2}^{N} F_p[z^{(p)}] + o(\|z\|^N) \,, \\ \varphi(x) = \sum_{p \geq 2}^{N} \varphi_p x^p + o(|x|^N) \,, \\ f(x) = \sum_{p \geq 2}^{N} a_p x^p + o(|x|^N) \,. \end{cases} \tag{24}$$

Differentiating (22) with respect to t and replacing $\frac{dz}{dt}$ by $F(z)$, dx/dt by $f(x)$, and identifying powers in x we obtain a hierarchy of equations of the form :

$$\begin{cases} L_-\varphi_2 - a_2\xi_0 + F_2[\xi_0^{(2)}] = 0 \,, \\ L_-\varphi_3 - a_3\xi_0 + 2F_2[\xi_0, \varphi_2] + F_3[\xi_0^{(3)}] - 2a_2\varphi_2 = 0 \,, \\ \ldots\ldots \\ L_-\varphi_p - a_p\xi_0 + R_p = 0 \,. \end{cases} \tag{25}$$

Since L_- acts in E_- , the compatibility condition leads to

$$(26)\quad \begin{cases} a_2 = (F_2[\xi_0^{(2)}] \mid \xi_0^*) , \text{ then } \varphi_2 = -L_-^{-1} P_- F_2[\xi_0^{(2)}] , \\ a_3 = (2 F_2[\xi_0, \varphi_2] + F_3[\xi_0^{(3)}] \mid \xi_0^*) , \\ \ldots\ldots \end{cases}$$

We may notice on this example that E_- is the subspace orthogonal to ξ_0^* , i.e we just apply Fredholm alternative to solve (25) (see § 1.4) .

2.2.3. Hopf bifurcation.

Before giving the general result on the best possible form of the reduced differential equation on $\mathcal{M}_0$, let us consider another simple case called, a little abusively , "Hopf bifurcation" (Andronov (1937) proved all these things in dimension 2 , while Hopf (1942) proved them in dimension $n \geq 2$) .

We have, for this case, two simple conjugated eigenvalues $\pm i\omega_0$ for the linear operator L . Let us then define in $\mathbb{C}^n$ the eigenvectors ζ_0 , ζ_0^* such that :

$$(27)\quad \begin{cases} L\zeta_0 = i\omega_0\zeta_0 , & L\bar{\zeta}_0 = -i\omega_0\bar{\zeta}_0 , \\ L^*\zeta_0^* = -i\omega_0\zeta_0^* , & L^*\bar{\zeta}_0^* = i\omega_0\bar{\zeta}_0^* , \\ (\zeta_0 \mid \zeta_0^*) = 1 , & (\bar{\zeta}_0 \mid \zeta_0^*) = 0 , \end{cases}$$

where the scalar product is the one of $\mathbb{C}^n$. Now the subspace E_0 is 2 dimensional : any x in E_0 may be written as

$$(28)\qquad x = A\zeta_0 + \bar{A}\bar{\zeta}_0 ,$$

where A is a complex amplitude. We are looking for Φ such that

(29) $$z = A\zeta_0 + \bar{A}\bar{\zeta}_0 + \Phi(A,\bar{A})$$

represents a center manifold $\mathcal{M}_0$, while the differential equation

(30) $$\frac{dA}{dt} = f(A,\bar{A})$$

describes the dynamics on it. It is not necessary to assume Φ to be orthogonal to $\{\zeta_0^*, \bar{\zeta}_0^*\}$. In fact, this allows us to make, a nonlinear change of variables on A, in such a way that (30) becomes the simplest possible. The simplest possible equation (30) is called the "normal form".

To make the computation of the expansions of Φ and f, let us introduce the following notations :

(31) $$\begin{cases} \Phi(A,\bar{A}) = \sum_{p+q\geq 2} \Phi_{pq} A^p \bar{A}^q \quad , \quad \Phi_{pq} \in \mathbb{C}^n , \\ f(A,\bar{A}) = \sum_{p+q\geq 1} a_{pq} A^p \bar{A}^q \quad , \quad a_{pq} \in \mathbb{C} . \end{cases}$$

Differentiating (29) with respect to t, and using (18), $(24)_1$, and (30), we obtain, by identification of powers of A and $\bar{A}$ at orders 1 and 2, equations of the form :

(32) $$L\zeta_0 = a_{10}\zeta_0 + \bar{a}_{01}\bar{\zeta}_0$$

(33) $$L\Phi_{20} + F_2[\zeta_0^{(2)}] = a_{20}\zeta_0 + \bar{a}_{02}\bar{\zeta}_0 + 2a_{10}\Phi_{20} + \bar{a}_{01}\Phi_{11}$$

(34) $$L\Phi_{11} + 2F_2[\zeta_0,\bar{\zeta}_0] = a_{11}\zeta_0 + \bar{a}_{11}\bar{\zeta}_0 + 2a_{01}\Phi_{20} + 2\bar{a}_{01}\Phi_{02} + (a_{10} + \bar{a}_{10})\Phi_{11} .$$

and since z is real we have $\Phi_{pq} = \bar{\Phi}_{qp}$. Now the equation (32) gives

(35) $$a_{10} = i\omega_0 , \quad a_{01} = 0 .$$

Looking at (33) and (34) we can observe that L and $L-2i\omega_0$ are invertible, so we can choose $a_{20}=a_{02}=a_{11}=0$, and finally :

$$(36)\qquad \begin{cases} \Phi_{20} = (2i\omega_0 - L)^{-1} F_2[\zeta_0^{(2)}] \quad , \quad \Phi_{02} = \overline{\Phi}_{20} \, , \\ \Phi_{11} = -2L^{-1} F_2[\zeta_0, \overline{\zeta}_0] . \end{cases}$$

The order 3 in $A, \overline{A}$ leads to the system :

$$(37)\qquad (L-3i\omega_0)\Phi_{30} + 2F_2[\zeta_0, \Phi_{20}] + F_3[\zeta_0^{(3)}] = a_{30}\zeta_0 + \overline{a}_{03}\overline{\zeta}_0 \, ,$$

$$(38)\qquad (L-i\omega_0)\Phi_{21} + 2F_2[\zeta_0,\Phi_{11}] + 2F_2[\overline{\zeta}_0,\Phi_{20}] + 3F_3[\zeta_0^{(2)},\overline{\zeta}_0] = a_{21}\zeta_0 + \\ + \overline{a}_{12}\overline{\zeta}_0 .$$

Since $L-3i\omega_0$ is invertible we can choose $a_{30}=a_{03}=0$, and Φ_{30} is deduced from (37) in a unique way. Now we observe that in (32) , the operator acting on Φ_{21} is not invertible. The Fredholm alternative applies (see 1.4) and gives

$$(39)\qquad a_{21} = \left(2F_2[\zeta_0,\Phi_{11}] + 2F_2[\overline{\zeta}_0,\Phi_{20}] + 3F_3[\zeta_0^{(2)},\overline{\zeta}_0] \,|\, \zeta_0^*\right),$$

while we can choose $a_{12}=0$. Now Φ_{21} is defined up to $\alpha\zeta_0$ (kernel of $L-i\omega_0$). We can fix its determination by posing

$$(40)\qquad (\Phi_{21} | \zeta_0^*) = 0 \; .$$

Finally we obtain an amplitude equation (30) of the form :

$$(41)\qquad \frac{dA}{dt} = A\left[i\omega_0 + a_{21}|A|^2 + a_{32}|A|^4 + \cdots\right]$$

which is <u>equivariant under rotations</u> of the complex plane, i.e if we change A into $Ae^{i\varphi}$ the equation is multiplied by $e^{i\varphi}$. We shall see that this is a special case of a very general result on "normal forms" .

Strictly speaking, we have not yet a Hopf bifurcation since we need a parameter μ such that for $\mu=0$ the above singularity holds.

Hence we consider in fact the system

(42) $$\frac{dz}{dt} = F(\mu, z) \ , \quad \mu \in \mathbb{R}, \quad z(t) \in \mathbb{R}^n,$$

with

(43) $$F(0,0) = 0 \quad , \quad D_z F(0,0) = L \, ,$$

and $\pm i\omega_0$ simple eigenvalues of L , other eigenvalues of L having a negative real part. Then, instead of (29) we set

(44) $$z = A\zeta_0 + \bar{A}\bar{\zeta}_0 + \Phi(\mu, A, \bar{A}),$$

which means that the center manifold now depends on μ , and we look for an amplitude equation

(45) $$\frac{dA}{dt} = f(\mu, A, \bar{A}).$$

We can make the same analysis as above, adding powers of μ in the expansions :

(46) $$f(\mu, A, \bar{A}) = \sum_{r+p+q \geq 2} \mu^r a_{rpq} A^p \bar{A}^q$$

We obtain the same system as (32) (33) (34) (37) (38) for $r = 0$:

(47) $$a_{010} = i\omega_0 \ , \quad a_{001} = a_{020} = a_{011} = a_{002} = a_{030} = a_{012} = a_{003} = 0 \, ,$$

and a_{021} given by (39). Now we also have :

(48) $$(L - i\omega_0)\Phi_{110} + F_{11}\zeta_0 = a_{110}\zeta_0 + \bar{a}_{101}\bar{\zeta}_0 \, ,$$

hence

(49) $$a_{110} = (F_{11}\zeta_0 \mid \zeta_0^*)$$

and we can choose $a_{101} = 0$. Finally we obtain, instead of (41) an amplitude equation such that

(50) $$\frac{dA}{dt} = (i\omega_0 + a\mu)A + bA|A|^2 + \text{h.o.t.}$$

which is equivariant under rotations in $\mathbb{C}$, up to an arbitrarily high order (if F is $\mathcal{C}^\infty$).

On the principal part of (50), we can easily show the existence of a periodic solution :

(51) $$A = \rho_0 e^{i\Omega t},$$

with

(52) $$\begin{cases} \rho_0 = \left(-\frac{a_r\mu}{b_r}\right)^{1/2} + O(|\mu|^{3/2}), \\ \Omega = \omega_0 + a_i\mu + b_i\rho^2 + \cdots \end{cases}$$

If we put (50) in polar form, it is also easy to study the stability of this periodic solution : if we set

(52) $$\rho = \rho_0 + \rho' \quad , \quad A = \rho e^{i\varphi}$$

then we obtain

(53) $$\begin{cases} \frac{d\rho'}{dt} = 2b_r\rho_0^2\rho' + \text{h.o.t.} \\ \frac{d\varphi}{dt} = \Omega + 2b_i\rho_0\rho' + \text{h.o.t.} \end{cases}$$

If $b_r < 0$, then the periodic solution bifurcates for $a_r\mu > 0$ and it is attractive, while if $b_r > 0$ it bifurcates for $a_r\mu < 0$ and it is repelling.

We have to take care of the fact that we truncated (50) to find these results. It can be proved that the existence result as well as the stability result persist when we consider the full equation (50). The basic fact is that we are in dimension 2 and that we can built a first return map on the real axis, on which we can look for a fixed point (one dimensional prblem). This problem has the structure of a pitchfork bifurcation such as in § 1.6

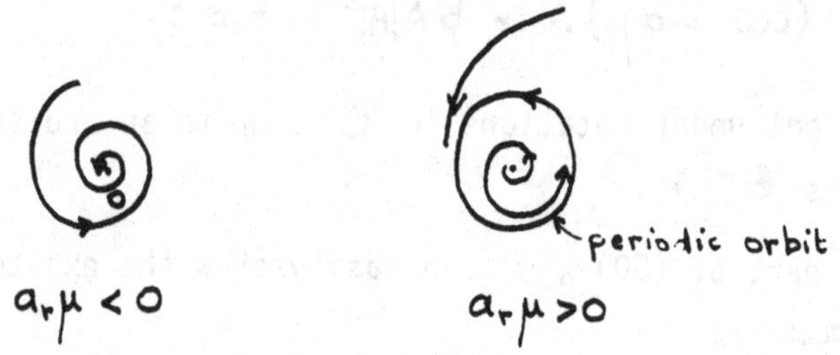

Figure 5 : Hopf bifurcation $(b_r < 0)$

2.2.4 . General computation of the amplitude equations.

Let us now consider a system obeying the differential equation in $\mathbb{R}^n$:

(54) $$\frac{dz}{dt} = F(\mu, z) \quad , \mu \in \mathbb{R} ,$$

where $F(0,0)=0$, $D_z F(0,0) = L$ satisfies the assumptions of §2.2.1 . Then the idea is to look for Φ and N such that

(55) $$\begin{cases} z = x + \Phi(\mu, x) & , x \in E_0 , \\ \Phi(\mu, x) = O(|\mu| + |x|^2), \end{cases}$$

and

(56) $$\frac{dx}{dt} = L_0 x + N(\mu, x) \quad , \quad N(0,0) = 0 \quad , D_x N(0,0) = 0 ,$$

with the simplest possible N . In fact we shall compute the Taylor expansions of Φ and N , hence the results are true up to an arbitrarily high order (not as a convergent series). The main result is the following [5] : polynomials Φ and N may be found such that

(57) $$N(\mu, e^{L_0^* t} x) = e^{L_0^* t} N(\mu, x) \quad , x \in E_0 , t \in \mathbb{R} ,$$

where L_0^* is the adjoint of L_0 in E_0 . We see that this gives

precisely the result in the Hopf bifurcation case since the group $\{e^{L_0^* t}\}_{t\in\mathbb{R}}$ is the rotation group in E_0. The equivariance of the normal form N under the group generated by the adjoint L_0^* may be used to compute a priori the form N , as we shall see below .
Let us introduce some notations to show the general method of computation for Φ and N :

$$(58)\quad \begin{cases} F(\mu,z) = \sum_{p+q\geq 1} \mu^p F_{pq}[z^{(q)}] , & F_{01}=L, z\in\mathbb{R}^n \\ \Phi(\mu,x) = \sum_{p+q\geq 1} \mu^p \Phi_{pq}[x^{(q)}] , & x\in E_0, \Phi_{01}=0, \\ N(\mu,x) = \sum_{p+q\geq 1} \mu^p N_{pq}[x^{(q)}] \in E_0 , & x\in E_0, N_{01}=0 . \end{cases}$$

Differentiating (55) with respect to t , and using (54) , (56) , we identify powers of μ and x to find a hierarchy of equations :

$$(59)\quad L\,\Phi_{10} = N_{10} - F_{10} ,$$

$$(60)\quad L\,\Phi_{02}[x^{(2)}] - D_x\Phi_{02}[x^{(2)}].L_0 x = N_{02}[x^{(2)}] - F_{02}[x^{(2)}] ,$$

$$(61)\quad L\,\Phi_{11}(x) - \Phi_{11}L_0 x = N_{11}(x) - F_{11}(x) - 2F_{02}[x,\Phi_{10}] + D_x\Phi_{02}[x^{(2)}]N_{10}$$

$$(62)\quad L\,\Phi_{20} = N_{20} - F_{20} - F_{11}\Phi_{10} - F_{02}[\Phi_{10}^{(2)}] + \Phi_{11}N_{10}$$

....

$$(63)\quad L\,\Phi_{pq}[x^{(q)}] - D_x\Phi_{pq}[x^{(q)}]L_0 x = N_{pq}[x^{(q)}] - R_{pq}[x^{(q)}] ,$$

where the unknown at each step are Φ_{pq} and N_{pq} , and where R_{pq} is a known function of $\Phi_{p'q'}$, $N_{p'q'}$ with $p'+q'\leq p+q$, $p'\leq p-1$.
To solve (59) , (62) or any of the equation for Φ_{p0} , N_{p0} , we can choose N_{p0} in $\ker L_0^*$ in E_0 in such a way that Fredholm alternative applies , i.e. the right hand sides of these equations are orthogonal to $\ker L_0^*$. Such N_{p0} are then defined

in a unique way by a standard algebraic procedure. In fact, all other equations (60) , (61) , (63) can be solved in the same way.
Let us split into two parts the equation (63) :

(64) $$L_- P_- \Phi_{pq}[x^{(q)}] - D_x P_- \Phi_{pq}[x^{(q)}] L_o x = - P_- R_{pq}[x^{(q)}] ,$$

(65) $$L_o P_o \Phi_{pq}[x^{(q)}] - D_x P_o \Phi_{pq}[x^{(q)}] L_o x = N_{pq}[x^{(q)}] - P_o R_{pq}[x^{(q)}] .$$

Equation (64) can be solved explicitely by :

(66) $$P_- \Phi_{pq}[x^{(q)}] = \int_0^\infty e^{L_- t} P_- R_{pq}[(e^{-L_o t} x)^{(q)}] dt ,$$

where the convergence of the integral is due to the exponential decay when $t \to \infty$ of $e^{L_- t}$.
Now , the equation (65) is a linear equation in $P_o \Phi_{pq}$ which is a homogeneous polynomial of degree q in x , taking values in E_o . This equation has the form :

(67) $$\mathcal{A}(P_o \Phi_{pq}) = N_{pq} - P_o R_{pq} ,$$

and, as usual, we can choose N_{pq} in the kernel of the adjoint of $\mathcal{A}$ to make the right hand side orthogonal to this kernel. The only problem is to define a scalar product and then to compute the adjoint of $\mathcal{A}$ to be able to compute its kernel! This problem was simply solved recently [5] , with an extension of the scalar product in the space of polynomials such that the differentiation is the adjoint of multiplication by the corresponding variable ! With this choice the adjoint $\mathcal{A}^*$ has the same form as $\mathcal{A}$ but with L_o^* instead of L_o. Hence we choose N_{pq} such that

(68) $$L_o^* N_{pq}[x^{(q)}] - D_x N_{pq}[x^{(q)}] L_o^* x = 0 \quad , \text{for any } (p,q) \neq (0,1).$$

In summing up we get

(69) $$L_0^* N(\mu,x) - D_x N(\mu,x)\cdot L_0^* x = 0$$

which is a linear partial differential equation, very useful to a priori compute the normal form N (see below) . In fact (69) is equivalent to (57) as it can be shown in differentiating (57) with respect to t .

2.2.5. Examples of amplitude equations.

Example 1. Assume that L has two pairs of simple pure imaginary eigenvalues $\pm i\omega_0$, $\pm i\omega_1$ and let us define

$$x = (A, \bar{A}, B, \bar{B}) .$$

The components of $N(\mu,x)$ are sums of monomials of the form $A^p \bar{A}^q B^r \bar{B}^s$. We first observe that

$$e^{L_0^* t} x = (e^{-i\omega_0 t} A, e^{i\omega_0 t} \bar{A}, e^{-i\omega_1 t} B, e^{i\omega_1 t} \bar{B}),$$

hence the property (57) leads, for the first component, to coefficients such that

$$\omega_0(p-q-1) + \omega_1(r-s) = 0 ,$$

and for the third component :

$$\omega_0(p-q) + \omega_1(r-s-1) = 0 .$$

As a consequence, we see that if ω_0/ω_1 is irrational, we only need to keep in the first component the coefficients such that $p=q+1$, $r=s$, and in the third one $r=s+1$, $p=q$. Now, if $\omega_0/\omega_1 = m/n$ (m and n have no common divisor) , then $p=q+1+kn$, $r=s-km$, $k \in \mathbb{Z}$ for the first component. Finally the normal form reads :

(70) $$\begin{cases} \dfrac{dA}{dt} = (i\omega_0 + a_0\mu)A + A\,P_0(|A|^2, |B|^2, A^n\bar{B}^m) + \bar{A}^{n-1}B^m P_1(|A|^2, |B|^2, \bar{A}^n B^m), \\[2ex] \dfrac{dB}{dt} = (i\omega_1 + a_1\nu)B + B\,Q_0(|A|^2, |B|^2, \bar{A}^n B^m) + A^n \bar{B}^{m-1} Q_1(|A|^2, |B|^2, A^n\bar{B}^m) , \end{cases}$$

where P_0, P_1, Q_0, Q_1 are polynomials in their arguments, and μ and ν are two complex parameters (it is necessary here to consider 2 real parameters to make the two pairs of eigenvalues on the imaginary axis simultaneously and another real parameter is useful for taking account of the "detuning" between ω_0 and ω_1 (the fourth parameter is a time rescaling). Notice that all the previous analysis works as well with an arbitrary number of parameters.

Example 2 . Assume that 0 is a double eigenvalue of L, such that

(71) $$L_0 = \begin{pmatrix} 0 & 1 \\ 0 & 0 \end{pmatrix}.$$

We define $x = (A, B)$ and $N = (N_1, N_2)$, then the partial differential system (69) may be written as :

(72) $$A \frac{\partial N_1}{\partial B} = 0, \quad A \frac{\partial N_2}{\partial B} = N_1 .$$

This leads immediately to the solution :

$$N_1(A,B) = A\varphi_1(A), \quad N_2(A,B) = B\varphi_1(A) + \varphi_2(A)$$

where φ_1 and φ_2 are polynomials (depending eventually on parameters). We then obtain amplitude equations of the form :

(73) $$\begin{cases} \dfrac{dA}{dt} = B + A\varphi_1(A) \\ \dfrac{dB}{dt} = B\varphi_1(A) + \varphi_2(A) . \end{cases}$$

hanging again variables it is easy to arrive to

(74) $$\begin{cases} \dfrac{dA}{dt} = B' , \\ \dfrac{dB'}{dt} = B'\varphi_1(A) + \varphi_2'(A) , \end{cases}$$

and if we introduce explicitely parameters, it is not hard to show that 2 parameters are generally sufficient such that the generic system is :

$$(75)\quad \begin{cases} \dfrac{dA}{dt} = B , \\ \dfrac{dB}{dt} = \mu_1 + \mu_2 B + AB\,P_1(\mu_1,\mu_2,A) + A^2 P_2(\mu_1,\mu_2,A) . \end{cases}$$

This system is the source of a great amount of analysis [6] .

2.3 . Presence of symmetries.

2.3.1. General result.

Let us assume that the system (54) is invariant under some symmetry. We denote by T a linear invertible operator in $\mathbb{R}^n$ such that

$$(76)\quad F(\mu, Tz) = T F(\mu, z) ,$$

for any μ and z . In differentiating (76) with respect to z at 0 , we observe that T commutes with L . Let us denote by T_0 the restriction of T to the subspace E_0 (since it is also invariant under T) , then the main result is as follows (see [5]) .

Assume that T_0 is a unitary operator on E_0, i.e.

$$(77)\quad T_0^{-1} = T_0^* ,$$

then the center manifold can be found such that

$$(78)\quad \Phi(\mu, T_0 x) = T\,\Phi(\mu, x) ,$$

and the amplitude equation such that

$$(79)\quad N(\mu, T_0 x) = T_0 N(\mu, x) .$$

2.3.2 . Pitchfork bifurcation

The simplest symmetry is given by a linear operator S such that

$$(80)\quad S^2 = Id \quad , \quad S \neq Id \quad ,$$

and $$F(\mu, Sz) = S\,F(\mu, z)\,.$$

Let us assume that 0 is a simple eigenvalue of L, other eigenvalues being of strictly negative real part. This is the case for instance in the example shown at §1.2 .

We denote by ξ_0 the eigenvector such that

$$L\xi_0 = 0$$

Then, since $S\xi_0$ is also an eigenvector for the same 0 eigenvalue, we have two possibilities (due to (80)) :

$$\text{either} \quad S\xi_0 = \xi_0 \quad , \text{ or } \quad S\xi_0 = -\xi_0 \quad .$$

(i) Let us first assume that $S\xi_0 = \xi_0$, then we have, thanks to (78):

$$Sz = S(x\xi_0) + S\,\Phi(\mu, x\xi_0)$$
$$= x\xi_0 + \Phi(\mu, x\xi_0)\,.$$

Hence the family of steady solutions, found for the amplitude equation

$$\frac{dx}{dt} = a\mu + bx^2 + \text{h.o.t.}$$

is invariant under S , i.e all steady solutions are symmetric.

(ii) Let us now assume that $S\xi_0 = -\xi_0$, then we have

$$S\,\Phi(\mu, x\xi_0) = \Phi(\mu, -x\xi_0)$$

and the amplitude equation

(81) $$\frac{dx}{dt} = n(\mu, x)$$

satisfies an oddness property (thanks to (79)) :

(82) $$n(\mu, -x) = -n(\mu, x)\,.$$

Hence $x = 0$ is a steady <u>persistent</u> solution for $\mu \neq 0$, which gives a <u>symmetric solution</u> for the system (54), since

$$Sz = S(0\xi_0) + S\,\Phi(\mu, 0\xi_0) = \Phi(\mu, 0)\,.$$

Now we have another family of steady solutions since (81) might be written as follows :

(83) $$\frac{dx}{dt} = a\mu x + bx^3 + \text{h.o.t}\,.$$

we obtain

(84) $$\mu = -\frac{b}{a}x^2 + \text{h.o.t.}$$

This is a "pitchfork bifurcation" of the same type as the one at example 1.2. It is easy to prove that the stabilities of the symmetric branch $(x=0)$ and of the other solution (84), which breaks the symmetry, are opposite.

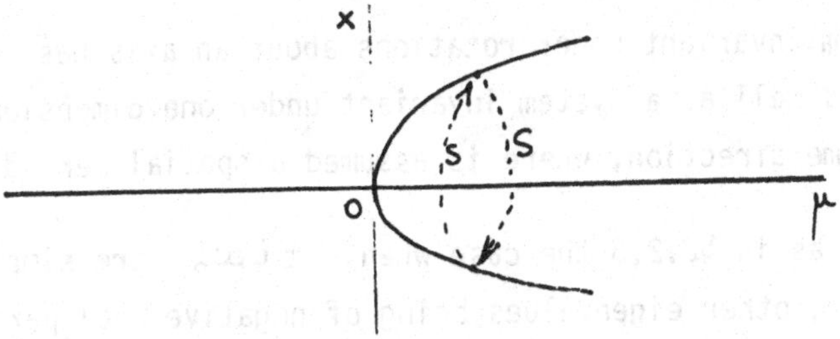

Figure 6 : Pitchfork bifurcation $(ab<0)$

Let us assume now that the system contains a little imperfection, i.e. we have

(85) $$\frac{dz}{dt} = F(\mu, z, \delta)$$

where for $\delta = 0$ the previous assumptions hold to obtain a pitchfork bifurcation. Then the normal form may be computed with this additional parameter as well. We know that for $\delta = 0$, the normal form is odd, hence we have in general :

(86) $$\frac{dx}{dt} = a\mu x + bx^3 + c\delta + \text{h.o.t.}$$

We then recover, for the steady solutions, the situation described in § 1.7.2, i.e. we break the pitchfork bifurcation (see Figure 9 of chapter 1).

2.3.3 . $SO(2)$ invariance.

A system (54) has a $SO(2)$ invariance if it commutes with a representation R_φ of the group of plane rotations. This means that we have a family R_φ, $\varphi \in \mathbb{R}$, of linear operators in $\mathbb{R}^n$ such that

$$(87)\quad \begin{cases} F(\mu, R_\varphi z) = R_\varphi F(\mu, z), \\ R_{\varphi_1} R_{\varphi_2} = R_{\varphi_1+\varphi_2}, \quad R_0 = R_{2\pi} = Id . \end{cases}$$

Any physical system invariant under rotations about an axis has this invariance, as well as a system invariant under one dimensional translations in some direction, where is assumed a spatial periodicity.

Let us consider as in §2.2.3 the case when $\pm i\omega_0$ are simple eigenvalues of L, other eigenvalues being of negative real part. We denote by ζ_0 the eigenvector belonging to $i\omega_0$. Then we have

$$(88)\quad R_\varphi \zeta_0 = e^{i\ell\varphi} \zeta_0$$

with some integer ℓ. This is a consequence of the fact that $R_\varphi \zeta_0$ is an eigenvector belonging to $i\omega_0$, and that we have the group properties (87) for R_φ.

If $\ell = 0$, the eigenvectors ζ_0 and $\bar{\zeta}_0$ are invariant under R_φ (they are "axisymmetric"). Then it can be deduced that the bifurcating solution which is self-oscillating in time, is itself axisymmetric (thanks to (78)).

The most interesting case is when $\ell \neq 0$ in (88). We have $x = A\zeta_0 + \bar{A}\bar{\zeta}_0$ (as in §2.2.3) and

$$R_\varphi(A\zeta_0 + \bar{A}\bar{\zeta}_0) = Ae^{i\ell\varphi}\zeta_0 + \overline{(Ae^{i\ell\varphi})}\,\bar{\zeta}_0 .$$

We computed the bifurcated self-oscillating solutions in §2.2.3 :

$$A = \rho_0 e^{i\Omega t},$$

hence R_φ does the same as changing t into $t + \ell\varphi/\Omega$:

$$R_\varphi x(t) = x(t + \ell\varphi/\Omega),$$

and if we come back to the solution z of (54) we obtain , thanks to (78) :

$$R_\varphi z(t) = z(t + \ell\varphi/\Omega) ,$$

which can be written in a simpler way :

$$z(t) = R_{\Omega t/\ell}\, z(0) . \tag{89}$$

This shows that these periodic solutions of (54) have a particular structure : they are called "rotating waves". When one rotates the angle φ by 2π , we obtain ℓ times the period (ℓ waves) .

Remark . We saw in §2.2.3 that the normal form of the complex amplitude equation is equivariant under rotation. This was proved in the general context, up to an arbitrarily fixed order. Here, with the $SO(2)$ symmetry, this property is valid for the full equation.

2.3.4 . $O(2)$ invariance - Travelling waves - Standing waves.

A system (54) has an $O(2)$ invariance if it commutes with i) a representation R_φ of the group of plane rotations and ii) with a symmetry S such that

$$R_\varphi S = S R_{-\varphi} . \tag{90}$$

A physical example with this symmetry can be for instance a case with a cylindrical geometry, with nothing breaking the symmetry $\theta \mapsto -\theta$ for the angle around the axis. Another very common case is a system invariant under translations along a direction oz and invariant under reflexions $z \mapsto -z$, when a spatial periodicity is prescribed.

The most interesting singular situation is when the eigenvalues are no longer simple. It is the general case here. Assume λ is an eigenvalue of L , with an eigenvector ζ . Then $S\zeta$ is also an eigenvector, and so $R_\varphi \zeta$ for any $\varphi \in \mathbb{R}$. Let us choose an eigenvector ζ which is also an eigenvector for R_φ (since R_φ and L commute), then we have

$$R_\varphi \zeta = e^{i\ell\varphi}\zeta \quad , \ \ell \text{ integer}.$$

Now

$$R_\varphi(S\zeta) = S R_{-\varphi}\zeta = e^{-i\ell\varphi}(S\zeta),$$

and this proves that if $\ell \neq 0$, $S\zeta$ is not colinear to ζ.

Let us now assume that $\pm i\omega_0$ are double eigenvalues of L, other eigenvalues being of strictly negative real part, and assume that S and R_φ are not trivial in the eigenspace belonging to $\pm i\omega_0$.

Then we have the decomposition

$$\begin{cases} x = A\zeta_0 + \bar{A}\bar{\zeta}_0 + B\zeta_1 + \bar{B}\bar{\zeta}_1 , \\ \zeta_1 = S\zeta_0 , \quad R_\varphi\zeta_0 = e^{i\ell\varphi}\zeta_0 , \quad R_\varphi\zeta_1 = e^{-i\ell\varphi}\zeta_1 . \end{cases}$$

The amplitude equations have the form

$$\begin{cases} \dfrac{dA}{dt} = f(\mu, A, \bar{A}, B, \bar{B}) , \\[2mm] \dfrac{dB}{dt} = g(\mu, A, \bar{A}, B, \bar{B}) , \end{cases}$$

where f and g satisfy (see (79) with $T_0 = R_\varphi$ and S) :

$$(91) \quad \begin{cases} f(\mu, e^{i\ell\varphi}A, e^{-i\ell\varphi}\bar{A}, e^{-i\ell\varphi}B, e^{i\ell\varphi}\bar{B}) = e^{i\ell\varphi} f(\mu, A, \bar{A}, B, \bar{B}) \\ f(\mu, B, \bar{B}, A, \bar{A}) = g(\mu, A, \bar{A}, B, \bar{B}) \end{cases}$$

Now the normal form (f, g) has to satisfy the property (57).

Here

$$e^{L_0^* t} \quad (A, B) \longmapsto (A e^{-i\omega_0 t}, B e^{-i\omega_0 t}),$$

hence we have :

(92) $f(\mu, e^{-i\omega_0 t}A, e^{i\omega_0 t}\bar{A}, e^{-i\omega_0 t}B, e^{i\omega_0 t}\bar{B}) = e^{-i\omega_0 t} f(\mu, A, \bar{A}, B, \bar{B}).$

It is not hard to show that (91)-(92) lead to a very simple form for (f,g) :

$$(93)\quad \begin{cases} \frac{dA}{dt} = (i\omega_0 + a\mu)A + b\,A|A|^2 + c\,A|B|^2 + h.o.t. \\ \frac{dB}{dt} = (i\omega_0 + a\mu)B + c\,B|A|^2 + b\,B|B|^2 + h.o.t. \end{cases}$$

The study of this system is made, in particular in [3] . The most interesting cases are when (i) $b_r < 0$, $b_r > c_r$ (assume $a_r > 0$), or when (ii) $b_r < c_r$, $b_r + c_r < 0$.

For $\mu > 0$, in the first case (i) we tends asymptotically in time to a solution of the form (or the symmetric solution) :

$$(94)\qquad A(t) = r_0 e^{i(\Omega_0 t + \varphi_0)}, \quad B(t) = 0,$$

where
$$\begin{cases} a_r\mu + b_r r_0^2 + \cdots = 0, \\ \Omega_0 = \omega_0 + a_i\mu + b_i r_0^2 + \cdots \end{cases}$$

This solution is a rotating wave as described in §2.3.3 , but it is often called a "travelling wave", due to the physical origin of the $O(2)$ symmetry if it comes from a translational invariance .

For $\mu > 0$, in the second case (ii) we tends asymptotically in time to a solution of the form :

$$(95)\qquad A(t) = r_0 e^{i(\Omega_1 t + \varphi_0)} \qquad B(t) = r_0 e^{i(\Omega_1 t + \varphi_1)}$$

where
$$\begin{cases} a_r\mu + (b_r + c_r) r_0^2 + \cdots = 0 \\ \Omega_1 = \omega_0 + a_i\mu + (b_i + c_i) r_0^2 + \cdots \end{cases}$$

This type of solutions is again periodic but of different kind, and we can write :

(96) $$z(t) = R_{\Omega_1 t/\ell}\, x_{\varphi_0} + R_{-\Omega_1 t/\ell}\, \tilde{x}_{\varphi_1} + \Phi(\mu, R_{\Omega_1 t/\ell}\, x_{\varphi_0} + R_{-\Omega_1 t/\ell}\, \tilde{x}_{\varphi_1}),$$

where $x_{\varphi_0} = r_0(e^{i\varphi_0}\zeta_0 + e^{-i\varphi_0}\bar{\zeta}_0)$, $\tilde{x}_{\varphi_1} = r_0(e^{i\varphi_1} S\zeta_0 + e^{-i\varphi_1}\overline{S\zeta_0})$.

If $\varphi_0 = \varphi_1$, we have, thanks to (78) ,

(97) $$S\, z(t) = z(t)\ .$$

Making a phase shift on the origin of time, we can always arrange the solution to satisfy (97) . This solution is a superposition of two travelling waves running in opposite directions. It is called a "standing wave".

If the conditions (i) or (ii) are not realised it may just happen that the standing waves or the travelling waves , even though they exist, may not be seen since they are unstable (see [3]) .

To end this chapter let us mention for the reader a very interesting physical problem - The Couette-Taylor problem - where a symmetry $O(2) \times SO(2)$ occurs (see [3]) . In this case the rotational invariance around the axis of cylinders gives the $SO(2)$ symmetry, while the translational invariance with periodicity condition and reflexional invariance gives the $O(2)$ symmmetry.

3. PERIODICALLY FORCED SYSTEMS

3.1 Periodic forcing far from a singularity

Let us consider the following differential equation in $\mathbb{R}^n$:

(1) $$\frac{dz}{dt} = F(z) + \delta G(z,t) \quad , \quad z(t) \in \mathbb{R}^n,$$

where $F(0) = 0$, $\delta \in \mathbb{R}$ is close to 0 , G is T-periodic in t . We assume here that the linear operator $L = D_z F(0)$ has no eigenvalue on the imaginary axis, then it is not difficult to show that (1) has a unique T-periodic solution in some neighborhood of 0 . In fact, we can compute an expansion of this solution of the form :

(2) $$z(t) = \sum_{p \geq 1}^{k} \delta^p z_p(t) + o(\delta^k)$$

Replacing (2) in (1) and identifying powers of δ we get (using the standard notations for the Taylor expansions of F and G with respect to z in a neighborhood of 0) :

(3) $$\frac{dz_1}{dt} = Lz_1 + G_0(t)$$

(4) $$\frac{dz_2}{dt} = Lz_2 + F_2[z_1^{(2)}] + G_1(t).z_1$$

(5) $$\frac{dz_3}{dt} = Lz_3 + 2F_2[z_1, z_2] + F_3[z_1^{(3)}] + G_1(t).z_2 + G_2(t)[z_1^{(2)}]$$

....

At any order, we have to solve an equation of the form

(6) $$\frac{dy}{dt} = Ly + f(t)$$

where f is a known T-periodic function, and y is the unknown T-periodic function in $\mathbb{R}^n$. By using Fourier analysis, it is easy to solve (6) :

$$(7)\quad \begin{cases} Y = \sum_{q\in\mathbb{Z}} Y_q e^{iq\omega t}, & f = \sum_{q\in\mathbb{Z}} f_q e^{iq\omega t}, \quad \omega = 2\pi/T, \\ Y_q = (iq\omega - L)^{-1} f_q, & \end{cases}$$

since $iq\omega - L$ is invertible. We might note that the solution (7) can be written into the form :

$$Y(t) = \int_{-\infty}^{t} e^{L(t-s)} f(s)\, ds$$

provided that all eigenvalues of L have a negative real part (there is an analogous expression when some eigenvalues have a positive real part).

We can observe on (7) that if L has pure imaginary eigenvalues, then we can still compute the T-periodic solution of (6) provided that we have no resonance, i.e. provided that no eigenvalue is a multiple of $i\omega$.

3.2 Periodic forcing near a singularity

We now consider a system satisfying the following :

$$(8)\quad \frac{dz}{dt} = F(\mu, z) + \delta G(\mu, z, t) \quad \text{in } \mathbb{R}^n,$$

where $F(0,0) = 0$, μ and $\delta \in \mathbb{R}$, G is T-periodic in t. We moreover assume that the linear operator

$$D_z F(0,0) = L$$

satisfies the assumptions of § 2.2.1., i.e. L has some eigenvalues on the imaginary axis, while all the others have a negative real part. We split the space $\mathbb{R}^n$ into $E_0 \oplus E_-$ as in §2.2 and define projections P_0, P_- on these invariant subspaces.

The idea is now to look for the normal form of this system reduced on some center manifold like in 2.2.4.

In fact, the idea is to look for Φ and N such that

$$(9)\quad z = x + \Phi(\mu, \delta, t, x), \quad x \in E_0,$$

Φ is T periodic in t, $|\Phi| = O(|\mu| + |\delta| + |x|^2)$, and

$$\frac{dx}{dt} = L_0 x + N(\mu, \delta, t, x) \quad \text{in } E_0, \tag{10}$$

where $N(0,0,t,0) = 0$, $D_x N(0,0,t,0) = 0$, N is T periodic in t , and independent of t for $\delta = 0$.

Equation (9) defines a manifold in the space $\mathbb{R}^n \times T^1$ where T^1 is the circle $\mathbb{R}/T\mathbb{Z}$, since we have the T periodicity in t . This manifold has the dimension of E_0 plus 1 and it lies in a neighborhood of $0 \times T^1$. The idea is, as before, to find the dynamics on this manifold, given by (10) , with the simplest possible N . Our main result is that we can obtain N such that

(11) $e^{L_0^* t} N(\mu, \delta, t, e^{-L_0^* t} x)$ <u>is independent of</u> $t \in \mathbb{R}$;

we shall see later several applications of such a simple property. We see that it generalizes in an immediate manner what we found in the case $\delta = 0$.

Let us first show how the computation of N can be derived. We first set:

$$F(\mu, z) = \sum_{p+q \geq 1} \mu^p F_{pq}[z^{(q)}] \quad , \quad F_{01} = L \quad , \quad z \in \mathbb{R}^n,$$

$$G(\mu, z, t) = \sum_{p+q \geq 0} \mu^p G_{pq}(t)[z^{(q)}] \; , \; G_{pq}(t) \;\; T\text{-periodic},$$

(12)

$$\Phi(\mu, \delta, t, x) = \sum_{p+q+r \geq 1} \mu^p \delta^q \Phi_{pqr}(t)[x^{(r)}] \; , \; x \in E_0 \; , \; \Phi_{001} = 0,$$

$$N(\mu, \delta, t, x) = \sum_{p+q+r \geq 1} \mu^p \delta^q N_{pqr}(t)[x^{(r)}] \in E_0 \; , \; N_{001} = 0,$$

and Φ_{p0r} and N_{p0r} independent of t .

Differentiating (9) with respect to t , and using (8) , (10) , (12) , we can identify powers of μ , δ and x to find a hierarchy of equations of the form :

$$L \Phi_{pqr}(t)[x^{(r)}] - D_x \Phi_{pqr}(t)[x^{(r)}] L_0 x - \frac{\partial \Phi_{pqr}(t)[x^{(r)}]}{\partial t} = N_{pqr}(t)[x^{(r)}] - R_{pqr}(t)[x^{(r)}]. \tag{13}$$

where the unknown at each step are $\Phi_{pqr}(t)$, $N_{pqr}(t)$ and where $R_{pqr}(t)$ is a known function of previously computed terms. Like in § 2.2.4. , we project (13) on E_0 and on E_- , using the fact that $N_{pqr}(t)[X^{(r)}] \in E_0$. We first obtain explicitely $P_- \Phi_{pqr}(t)[x^{(r)}]$:

$$(14) \qquad P_- \Phi_{pqr}(t)[x^{(r)}] = -\int_0^\infty e^{L_- s} P_- R_{pqr}(t-s)\left[(e^{-L_0 s} x)^{(r)}\right] ds$$

after a simple multiplication of (13) by $e^{L_- s}$, and replacement of the argument x by $e^{-L_0 s} x$, and integration over $(0, \infty)$.
The real problem comes with $P_0 \Phi_{pqr}(t)[x^{(r)}]$. We already saw this problem for $q=0$ since then Φ_{p0r} and N_{p0r} are the same as in § 2.2.4. (independent of t). Using the same notation we can write :

$$(15) \qquad \mathcal{A}(P_0 \Phi_{pqr}(t)) - \frac{\partial}{\partial t}(P_0 \Phi_{pqr}(t)) = N_{pqr}(t) - P_0 R_{pqr}(t)$$

where $\mathcal{A}$ is the linear operator we studied in § 2.2.4. Using Fourier series for all these T-periodic functions of t we then obtain the infinite system :

$$(16) \qquad (\mathcal{A} - im\omega) P_0 \Phi_{pqr}^{(m)} = N_{pqr}^{(m)} - P_0 R_{pqr}^{(m)} , \quad m \in \mathbb{Z} .$$

We want to solve (16) with respect to $P_0 \Phi_{pqr}^{(m)}$ for the simplest possible $N_{pqr}^{(m)}$. This is the same type of problem than the one we solved in § 2.2.4. We can choose $N_{pqr}^{(m)}$ in the kernel of the adjoint operator :

$$(17) \qquad L_0^* N_{pqr}^{(m)}[X^{(r)}] - D_x N_{pqr}^{(m)}[X^{(r)}] L_0^* x + im\omega N_{pqr}^{(m)}[X^{(r)}] = 0$$

for any $m \in \mathbb{Z}$, $(p,q,r) \neq (0,0,1)$.

Summing up (17) for all p, q, r we obtain

$$(18) \qquad L_0^* N^{(m)}(\mu, \delta, x) - D_x N^{(m)}(\mu, \delta, x) . L_0^* x + im\omega N^{(m)}(\mu, \delta, x) = 0 .$$

Now multiplying by $e^{im\omega t}$ and summing up for all m , we finally obtain :

(19) $$\frac{d}{dt}\left[e^{L_0^* t} N(\mu,\delta,t,e^{-L_0^* t}x)\right]=0$$

which is (11) .

Of course, this is a global result for giving a priori a form for N . For a given problem we should have to i) look for the a priori form of N , using (11) , and ii) identify the terms of N in computing at each step the solution of (16) . Let us give some examples in the following § .

3.3 Periodic forcing of a saddle-node bifurcation

In such a case E_0 has dimension 1 and we have $x = A\xi_0$. The linear operator L_0 is 0 , hence $\mathcal{A}$ is 0 . It results that for any $m \neq 0$, (16) is solvable with $N^{(m)}_{pqr} = 0$. Hence the normal form N will be autonomous. Notice that this result will be true each time the only eigenvalue of L_0 is 0 , since then 0 will be the only eigenvalue of $\mathcal{A}$.

The result in the present case is that the amplitude equation reduces to :

(20) $$\frac{dA}{dt} = a\mu + b\delta + cA^2 + \text{h.o.t.}$$

We then recover the imperfect saddle-node bifurcation, with the additional property that each steady solution of (20) corresponds to a T-periodic solution of (8) (for $\delta \neq 0$). So, we have in fact a saddle-node of periodic solutions.Their stabilities are completely determined by (20) hence the result is the same as in the unforced case.

3.4 Periodic forcing of a Hopf-bifurcation

In such a case E_0 is 2 dimensional and we have

(21) $$x = A\zeta_0 + \bar{A}\bar{\zeta}_0 ,$$

with $L_0\zeta_0 = i\omega_0\zeta_0$. We observe that :

(22) $$e^{L_0^* t} : A \mapsto Ae^{-i\omega_0 t} ,$$

hence if we denote by $n(\mu,\delta,t,A,\bar{A})$, $\bar{n}(\dots)$ the components of N the property (11) leads to :

(23) $$\frac{d}{dt}\, e^{-i\omega_0 t} n(\mu,\delta,t,Ae^{i\omega_0 t},\bar{A}e^{-i\omega_0 t}) = 0 ,$$

where η is T-periodic in t. Denoting by $\eta_{pq}^{(m)}$ the coefficient of $A^p \bar{A}^q e^{mi\omega t}$, $\omega = 2\pi/T$, we obtain :

$$\frac{d}{dt} \eta_{pq}^{(m)} e^{[mi\omega + i\omega_0(p-q-1)]t} = 0,$$

hence the only remaining terms are the ones such that

(24) $$\omega_0(p-q-1) + m\omega = 0 .$$

If ω_0/ω is irrational then the only remaining terms in N are those with $m = 0$, $p = q+1$. This is the classical normal form for a Hopf bifurcation ! since the equation in A will be autonomous. In fact, we cannot verify such a property, we necessarily obtain some rational value for ω_0/ω each time we compute these frequencies. So, it is much more suitable to assume that ω_0/ω is very close to some rational number r/s with a not too large s, to be sure that all the little neighborhood of r/s does not contain another rational number with a lower s.

Hence, we assume that

(25) $$\omega_0/\omega = r/s + \gamma,$$

where γ is called the " detuning parameter ", and is assumed to be close to 0. Let us first look at the situation for $\gamma = 0$. The equation (24) is easily solvable : we have

(26)
$$p = q+1+\ell s, \quad m = -\ell r \quad \text{for } \ell \geq 0, \text{ or}$$
$$q = p-1+\ell s, \quad m = \ell r \quad \text{for } \ell > 0 .$$

This leads to terms in N of the form :

(27) $$|A|^{2q} A^{\ell s+1} e^{-\ell r i\omega t}, \quad |A|^{2p} \bar{A}^{\ell s-1} e^{\ell r i\omega t},$$

so, the lowest order term which depends on t is $\bar{A}^{s-1} e^{ri\omega t}$, and it has to occur in general, in factor of δ (for $\delta = 0$, all is autonomous).

Let us write the principal part of the amplitude equation in this resonant case :

(28) $$\frac{dA}{dt} = (i\omega_0 + a\mu) A + bA|A|^2 + c\delta \bar{A}^{s-1} e^{ri\omega t}.$$

We did not write all terms of the form $A|A|^{2p}$ since they are small compared to $bA|A|^2$ for $p>1$. Observe that here $r\omega = s\omega_0$, and that the equation (28) is equivariant under rotations of $2\pi/s$ in $\mathbb{C}$: multiply A by $e^{2i\pi/s}$ is the same as multiplying (28) by $e^{2i\pi/s}$.

If we now consider the case when the detuning parameter γ is $\neq 0$, we can say that ω_0 is close to $\tilde{\omega}_0 = r\omega/s$ and make the previous analysis with $\tilde{\omega}_0$ instead of ω_0, introducing γ as an additional parameter. Finally the normal form for $\gamma \neq 0$ is the same as (28) where $\omega_0 - \tilde{\omega}_0$ is close to zero.

Let us end this study by looking for a special kind of periodic solutions of (8) which have period $sT(\sim 2\pi r/\omega_0)$ due to the interaction with the natural frequency ω_0 (" frequency locking ").

We start with the amplitude equation (28) and make the following change of variables :

(29) $$A = B e^{i\frac{r}{s}\omega t}$$

The equation now becomes :

(30) $$\frac{dB}{dt} = [i(\omega_0 - \tfrac{r}{s}\omega) + a\mu]B + bB|B|^2 + c\delta \bar{B}^{s-1},$$

i.e. t disappears and the linear coefficient is small : $i\omega\gamma + a\mu$.
So, we can look for steady solutions of (30), using polar coordinates.

For $s \geqslant 2$, we observe that $B=0$ is a solution of (30). This corresponds to a T-periodic solution $z = \Phi(\mu,\delta,t,0)$ for the system (8). We recover the solution predicted at § 1, since ω_0 is not close a multiple of the forced frequency ω_0 (r and s have no common divisor). Now we write $B = \rho e^{i\varphi}$, and eliminating the solution $\rho = 0$, we have :

(31) $$\begin{cases} a_r\mu + b_r\rho^2 + \delta\rho^{s-2}(ce^{-si\varphi})_r = 0, \\ \omega\gamma + a_i\mu + b_i\rho^2 + \delta\rho^{s-2}(ce^{-si\varphi})_i = 0. \end{cases}$$

Using ρ as a parameter for the family of solutions :

$$(32)\quad \begin{cases} \mu = -\dfrac{b_r\rho^2}{a_r} - \dfrac{\delta\rho^{s-2}}{a_r}(ce^{-si\varphi})_r \,, \\ \omega\gamma + (b_i - b_r a_i/a_r)\rho^2 + \rho^{s-2}\delta[\alpha\cos(s\varphi) + \beta\sin(s\varphi)] = 0, \end{cases}$$

where

$$\alpha\cos(s\varphi) + \beta\sin(s\varphi) = (ce^{-si\varphi})_i - \frac{a_i}{a_r}(ce^{-si\varphi})_r \,.$$

Equation $(32)_2$ has solutions in φ , provided that

$$(33)\quad \left|\omega\gamma + (b_i - b_r a_i/a_r)\rho^2\right| < \delta\rho^{s-2}|\alpha||c|/|a_r|$$

In the space of parameters $(\mu, \delta, \omega_o/\omega)$ this region (taking account of $(32)_1$) may be represented by a horn, as indicated on Figure 1 . Into this " resonance horn " we have two families of s steady solutions

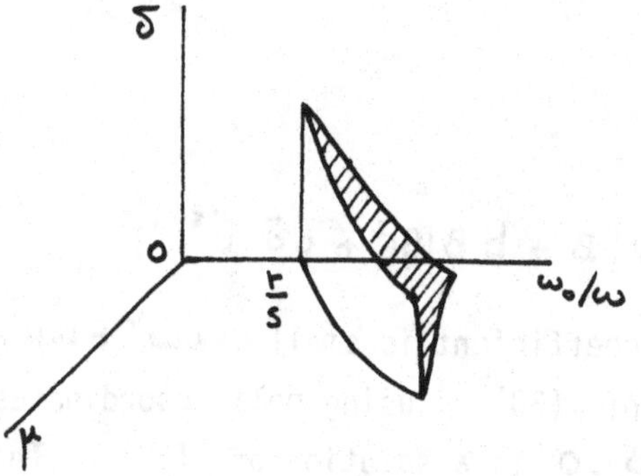

Figure 1

for (31) . For each family, we have

$$\varphi_k = \varphi_o + k\frac{2\pi}{s} \;,\; k = 0,1,\dots s-1 \,.$$

It can be easily shown that if $b_r < 0$, and δ is small enough, their stabilities are opposite. Each family corresponds to a periodic solution in A for the amplitude equation (28), of period sT (close to $2\pi r/\omega_o$). When we look at one of these solutions after a time T , this corresponds to a shift in φ by $2\pi r/s$

It has to be noticed that, if we take into account all terms in the amplitude equation (28) . , then the result still holds. This is the " frequency locking " phenomenon.

If δ is very small, it could be shown that there exists an invariant closed curve for (30) and that the steady solutions which are obtained $(B \neq 0)$ lie on this curve. This curve leads to an invariant torus for the original system in z , and in the frequency locking situation the trajectories are spiraling on the torus towards the stable sT-periodic solution.

For $s = 1$, then ω_o is close to a multiple of ω and

(31) cannot be written in the same way. It is the strongest resonance. Nevertheless, for a very small δ $(O(\rho^2))$ the same type of proof may be used to show the existence of two T-periodic solutions with opposite stabilities.

For a complete description of the dynamics in all cases, see the paper by J.M. GAMBAUDO [2]

BIBLIOGRAPHY

[1] V. ARNOLD, Chapitres supplémentaires de la théorie des équations différentielles ordinaires. MIR, Moscou, 1980

[2] J.M. GAMBAUDO, Perturbation of a Hopf bifurcation by an external time-periodic forcing. J. Differential Equations. 57, 179-199, 1985

[3] P. CHOSSAT, G. IOOSS, Primary and Secondary Bifurcations in the Couette-Taylor problem. Japan J. Appl. Math. 2, 1, 37-68, 1985

[4] R. COURANT, D. HILBERT, Methods of Mathematical Physics, Vol. 1. Interscience Pub. 1953

[5] C. ELPHICK, E. TIRAPEGUI, M. BRACHET, P. COULLET, G. IOOSS, A simple global characterization for normal forms of singular vector fields. Preprint n° 109 Université de Nice, 1986

[6] J. GUCKENHEIMER, P. HOLMES, Nonlinear oscillations, Dynamical systems and Bifurcations of vector fields. Applied Maths Sci. 42, Springer, 1983

[7] G. IOOSS, D.D. JOSEPH, Elementary stability and Bifurcation theory, U.T.M., Springer-Verlag, 1980

[8] T. KATO, Perturbation theory for linear operators. Springer Verlag, 1966.

Printed in the United States
By Bookmasters